FIDIC
合同条件准确应用指南
——2017年第2版重要条款翻译辨析

卢有杰 著

中国建筑工业出版社

图书在版编目（CIP）数据

FIDIC 合同条件准确应用指南：2017年第2版重要条款翻译辨析/卢有杰著. —北京：中国建筑工业出版社，2019.12
ISBN 978-7-112-24248-1

Ⅰ.①F… Ⅱ.①卢… Ⅲ.①建筑施工－经济合同－研究 Ⅳ.①TU723.1

中国版本图书馆CIP数据核字（2019）第217783号

责任编辑：李春敏　杨　杰
责任校对：张惠雯

FIDIC合同条件准确应用指南——2017年第2版重要条款翻译辨析
卢有杰　著
*
中国建筑工业出版社出版、发行（北京海淀三里河路9号）
各地新华书店、建筑书店经销
北京锋尚制版有限公司制版
北京建筑工业印刷厂印刷
*

开本：787×1092毫米　1/16　印张：6　字数：125千字
2019年11月第一版　　2019年11月第一次印刷
定价：39.00元
ISBN 978-7-112-24248-1
（34763）

版权所有　翻印必究
如有印装质量问题，可寄本社退换
（邮政编码100037）

前言

一、经济与社会背景

第二次世界大战结束以来，世界各国，特别是发达国家的政府工程，数目和规模日增，用途、功能、组成与形式亦随之日趋复杂。

另一方面，各国的建筑业，也发展迅速，技术日益提高，内部的纵向与横向分工和部门划分益加细密，为客户服务的能力更是竞相增强。工程承包企业彼此之间的竞争日益激烈，这种竞争的结果，就是在建筑业和国际建筑市场上出现了多种形式的部门、地区和企业垄断。

世界各地当前工程采购与交付的建筑市场中，民间工程同几十年前相比，虽然已经有了令人瞩目的长足发展，但是，以政府工程为主的买方垄断仍然是决定建筑业和国际建筑市场产业结构与市场竞争的决定性因素。从工程采购与交付双方的角度来看，国际建筑市场是买方市场，即买方的选择远远多于卖方；从交付方，即各国建筑业内部纵向各行业的角度来看，亦是买方市场，即总承包的选择远远多于专业分包、劳务分包和材料与设备供应厂家。

此外，自从20世纪70年代以来，世界各国金融市场日益开放，资金在各国之间的流动障碍越来越小，对外投资的机会和兴趣日益增加。工程所需资金，比以往任何时候都更容易取得。这种便利不但有利于工程采购方，对于工程的交付方亦为如此。

在经济全球化的背景下，世界各国，特别是发达国家的政府工程采购方式一直处于变动与发展过程之中，今天，已经达到了令人眼花缭乱的程度。但是，若拨开工程采购中日益繁复的各种合同关系的外观，就会看到主导这些采购方式与合同关系的总趋势，那就是：缩短工期、减少开支、确保质量、提高效益、保护环境，追求人类持续长久的发展。

自20世纪80年代以来，设计施工、设计采购施工、BOT、PPP，在政府工程采购中使用得越来越普遍，在民间工程中，亦已有实践。然而，施工服务，仍是政府工程和民间工程的主要采购对象。即使其他采购方式，在绝大多数情况下，经过多重分解，其基本合同仍然是施工合同。在不同的工程采购安排中，各种服务的提供者，都可以随时动员、调动和组织建筑和金融市场上的丰富资源，为工程最后采购方服务。

值得重申的是，现在的工程采购，特别是政府工程，很多是都利用卖方信贷，也就是由建筑企业为工程采购方垫付部分或全部资金，BOT和PPP即为如此。很多设计施工和设计采购施工的采购，也要由工程交付方垫付资金。我国的一些大型中央建筑企业为某些国家政府兴修大型基础设施，就是这样。

采购方式不同，双方的义务、权利和责任也就不同。因而，反映各种不同采购方式中双方权利、义务与责任的合同文件的内容、组成和编排也就不同。

另一方面，社会公众对政府工程的注意与关切日益增强，关心这些工程同社会、环境和自身的各种利害关系，经常对其寄托多方面的殷切期望。因此，工程的直接参与者，都负有满足社会公众期望的责任。在工程整个生命期内，直接参与者必须承受社会问责，都应随时回答、解释，并解决社会公众提出的各种疑问、批评和责难。

必须指出，FIDIC（国际咨询工程师联合会）从20世纪80年代起，就逐渐同世界银行、亚洲开发银行、非洲开发银行与其他一些国际金融组织结成了合作伙伴。众所周知，世界银行、亚洲开发银行，以及2015年12月由中国政府倡导在北京成立的亚洲基础设施投资银行等放贷对象都是各国政府，要求借款国政府在采购工程及其有关服务时，使用他们根据FIDIC合同文件"协调"后的版本。由此可知，FIDIC合同文件适用的是政府工程，若用于民间工程，则须做大的改动，改动的代价是巨大的。

这一点，也是我们阅读、理解和翻译FIDIC合同文件时，必须牢记的。

二、FIDIC合同文件法律背景

了解以下情况，有助于深刻理解FIDIC合同文件的概念、术语与含义。

第二次世界大战后，世界上许多专业团体都模仿ICE（英国土木工程师学会）的合同格式，编写出适合本国和本地区法律和术语的合同条件。但是，ICE格式主要是为英国国内使用而编写的。ACE（英国咨询工程师联合会）根据当时国际建筑市场的紧迫需要，经过ICE的同意，联合英国的建筑业出口集团动手编写了一份适合世界其他地方用的合同文件，并于1956年8月出版，一般叫作海外（土木工程）合同条件，简称ACE格式，是第一份国际土木工程施工标准合同条件，共分两部分，第一部分是68条的一般条件，第二部分是特别条件，以便顾及具体工程的具体情况，用于修改、补充第一部分。

在这之后不久，FIDIC和国际房屋建筑和公共工程联合会（后来改称欧洲国际建筑联合会（FIEC））在ACE格式的基础上于1957年8月出版了土木工程施工合同条件（国际）（第1版）（俗称"红皮书"），常称为FIDIC条件，用得很广泛。

由此可见，FIDIC合同条件中的合同概念源于英国普通法，措辞遵循这一法律。了解世界上主要的法律体系同FIDIC合同文件法律概念之间的基本关系，有助于更好地理解使用FIDIC合同文件时可能遇到的各种问题。

三、FIDIC对完善的追求

从FIDIC于1977年出版土木工程施工合同条件（国际）（第2版）开始，我国建筑企业的管理人员、工程咨询机构的专家、学校的教师就将FIDIC各个时期出版的所有合同文件及时地译成了汉文。屈指一算，这些工作已做了近40年矣。

从结识FIDIC以来，我们发现，FIDIC特别注意吸取自己所编写的合同文件的使用者，以及其他方面的批评和建议，及时将其纳入随后编写的各种版本之中。FIDIC对于新版本的措辞，高度重视含义清楚、准确，避免误解。随时淘汰含义模糊、易生歧义的词语、术语和表达方式。

当然，FIDIC也随着建筑业、咨询业，以及工程采购方式的进步和演变，引入一些新概念、新术语。

例如，在1957年出版的《土木工程施工合同条件（国际）（第1版）》中，使用maintenance period表示雇主接收工程后，承包人对于因其施工不当造成的缺陷负责修补的期间。然而，maintenance不能区分工程缺陷是何人所造成。所以，到了1977年的第2版，就将其改成了defects liability period。这一改动，虽然明确了承包人仅对因其施工不当造成的缺陷负责，但雇主还是希望承包人修补所有的工程缺陷。因此，在1999年的多种版本中将其改成了defects notification period。

再举一例。对权利得失的程度和范围的表达。

FIDIC在1987年第4版《土木工程施工合同条件》中，仅在第65.5款中用to the extent that限制了承包人要求支付款项的权利，即Save to the extent that the Contractor is entitled to payment under any other provision of the Contract, the Employer shall repay to the Contractor any costs of the execution of the Works……。

但是，12年后，1999年各种版本的许多条文都用了to the extent that限制合同双方的权利、义务的范围和程度。这种限制，使得合同双方的权利、义务与风险的分配更易为双方所接受，因而更加公平、合理。

以上两例可以说明，过去的几十年，FIDIC在编写合同文件时，坚持不断改进、完善，不固步自封，而是与时俱进，使其产品跟上工程采购方式前进的步伐。

四、本书的目的

FIDIC追求完美的精神值得我们学习。FIDIC在其出版物中，用词十分谨慎。很多都经过有丰富经验的律师的审核、推敲和润色。

我们在将其译成汉文时，没有理由不秉持同样的敬业精神，没有理由不慎而又慎，不可屈服于目前社会上含义模糊的流行词语和用法，更不能固守陈旧、狭隘和短视，应当大胆摒弃以往译文中的错误、片面、含糊或误导读者之处，在认真、准确地理解原文的基础上，拿出准确、通俗、易懂，令人耳目一新的译文，贡献给读者。

本书作者从1992年开始阅读和学习FIDIC合同文件各种版本及相应的汉译本。随着时间推移，作者益发感觉汉译本中有些地方值得商榷，分别于1996年与2011年写过两篇文章，探讨当时时已有汉译本中的若干问题。

遗憾的是，直到今天，有些不恰当的译法仍然在各种场合的讲话、文章、讨论中常常出现和误用。

作者希望趁着2017年第2版在我国尚未广泛传播之时，再次谈谈对迄今为止一些汉译的看法。

作者在本书中以EPC/交钥匙工程合同条件2017年第2版（以下简称"银皮书2017年第2版"）若干条文为例，介绍自己的译文，希望抛砖引玉，引起读者的兴趣，各抒己见，在批评和争论中集思广益，找到更好的词语和表达方式，希望作者与广大读者的共同努力有助于2017年第2版FIDIC合同文件将来的正式翻译超过以往所有的版本，使得广大的工程合同管理人员和其他有兴趣者能够准确、全面地理解和使用之。

对以往FIDIC合同文件译文中的某些问题产生的原因，作者也谈了自己不成熟的看法。

希望读者指出本书中的不当、错误、片面之处，作者将不胜感谢。

卢有杰
己亥年中秋，识于清华园荷清苑

目录

一　引言／001

二　银皮书2017年第2版的汉译／005

　　（一）DISCLAIMER 不担责声明／006

　　（二）NOTES 几点说明／006

　　（三）General Conditions 一般条件／006

　　（四）Advisory Notes to Users of FIDIC Contracts Where the Project Uses Building Information Modelling Systems 对项目使用建筑信息模型的FIDIC合同用户的若干建议说明／074

　　（五）NOTES ON THE PREPARATION OF SPECIAL PROVISIONS 编写特殊条文应注意之处／077

　　（六）Annex A EXAMPLE FORM OF PARENT COMPANY GUARANTEE 母公司保证书格式样本／079

　　（七）CONTRACT AGREEMENT 合同协议书／079

三　结束语／083

参考文献／085

一　引言

FIDIC合同条件在我国译成汉文，已经30多年。最早是1977年土木工程施工合同条件第3版[1]，继而是1987年第4版，1990年委托人/咨询工程师标准咨询服务协议书，接着是1995年设计-建造与交钥匙工程合同条件第1版，再就是施工合同条件、EPC/交钥匙工程合同条件和生产设备和设计-施工合同条件1999年第1版，2017年修改了这三本书1999年第1版，出版了第2版。

从1992年开始，笔者在阅读上述各版本及汉译本时，感觉以上各汉译本中有些地方值得商榷，分别于1996年[2]与2011年[3]写过两篇文章，探讨当时已有汉译本中的若干问题。

后来发现，直到今天，有些不恰当的译法仍然在讲话、文章、讨论等场合经常出现和使用。

这些问题主要表现在多义词择义不当、沿袭多年但不合道理的词语仍在使用、含义不同的英文词译成同一个汉语词汇等。

多义词择义不当，例如provide、cover；

沿袭多年但不合道理的词语，如claim。

含义不同的英文词译成同一个汉语词汇，如responsibility（responsible）、liability（liable）和obligation。

再如，将consent（应允）同agree一样都译成了"同意"，抹杀了consent同agree原意的区别。

还有，将同一个英文词accept（acceptable）在上下文没有多大区别的情况下，有时译成"接受"，有时译成"认可"。

目前，趁着2017年第2版还没有在我国广泛传播，笔者打算再次谈谈对迄今为止一些汉译的看法。更重要的是，打算对一些流传已久，但尚未有人公开表示异议的译法谈一谈笔者的看法，例如claim（索赔）、liable（负责）等。

另一方面，笔者也注意到，在这30多年当中，FIDIC也在不断地改进，随时改换含义不清，易生误解的英文词语。当然，也随着建筑业、咨询业，以及工程采购方式的进步和演变，引入一些新概念、新术语。

FIDIC在其出版物中，用词十分谨慎。很多都经过有丰富经验的律师的审核、推敲和润色。我们在将其译成汉文时，没有理由不慎而又慎，不可屈服于社会上含义模糊的流行词语和用法。

鉴于此，笔者极力主张，在以后的译文中摒弃那些沿袭已久，但不合理的词语，例如，索赔等。不应当以由来已久，已成习惯为借口，继续以讹传讹，贻笑后人。希望以后的译文得到较大改进。

本文以EPC/交钥匙工程合同条件2017年第2版（以下简称"银皮书2017年第2版"）为主，并按其顺序，介绍笔者的翻译，也顺便谈一谈对以往一些译法的看法。

最后，笔者对以往译文中的一些问题产生的原因，也谈了自己不成熟的看法。所有这些，希望读者指出其中不当、错误、片面之处，并随时告诉笔者，笔者不胜感谢之至。

二 银皮书 2017 年第 2 版的汉译

对于银皮书2017年第2版，本段分别指出各个部分翻译时应注意的问题。（一）～（七）是本书各个部分的编号，便于读者查找、对照，编号后面是这些部分的标题。

（一）DISCLAIMER 不担责声明

FIDIC publications are not exhaustive and are only intended to provide general guidance. They should not be relied upon in a specific situation or issue.（FIDIC出版物并非包罗一切，仅打算给予一般的指引。不应靠其处理具体情况或解决具体问题。）

【译注：这里的provide不是"提供"的意思。provide v.t. to afford or yield（出产；产生；成为；给予）[12]】

（二）NOTES 几点说明

For example, this edition provides:

例如，本版添置了：

1) greater detail and clarity on the requirements for notices and other communications;

1）对于通知和其他沟通讯息更详细、更清楚的要求；

2) provisions to address Employers' and Contractors' claims treated equally and separated from disputes;

2）将雇主与承包人的索要同争议分开而又同等处理的条文；……

【译注：这里，provide v.t.是make ready, do what is necessary for（准备、设置、设立、筹备）的意思。经常有人将provide译成"提供"。若不细想，译成"提供"也不觉得有什么不妥。后面还有许多例子，provide不能译成"提供"。】

（三）General Conditions 一般条件

本段依照银皮书2017年第2版一般条件的条款编号顺序行文。

1.1 Definitions

1.1.3 "Claim" means a request or assertion by one Party to the other Party for an entitlement or relief under any Clause of these Conditions or otherwise in connection with, or arising out of, the Contract or the execution of the Works.

【译注:"索要"指合同一方根据本条件任何条文或根据与本合同或实施本工程有关或由此产生的其他理由向另一方要求或坚持要求某项权利或补偿。

在FIDIC以前各版本的合同条件中,没有为claim下定义。结果,经常造成误解。为此,银皮书2017年第2版第一次为claim下了定义。

定义中的relief,若是查常见的英汉词典,例如文献[14],relief的释义有:减轻、解除、免除、宽慰;救济、救济品,等[14]在FIDIC以前各版本合同条件中,都将其译成"免除"。文献[16]150页将这里的relief译成了"救济"。《现代汉语词典》救济:用钱或物资帮助灾区或生活困难的人。但是,relief还有别的意思。relief n. legal remedy or redress(法律上的补救办法或补偿)。[12]

再看Claim,常有人译成"索赔",FIDIC以前各版本合同条件亦如此。《新华词典》"索赔"的释义是"国际贸易业务的一方违反合同的规定,直接或间接地给另一方造成损害,受损方向违约方提出损害赔偿要求。"

"赔"在汉语中有"补偿损失"、"亏耗"和"道歉或认错"之意。英文动词claim意思有:to ask for esp. as a right(因权利而要求);to take as the rightful owner(以正当所有者理由拿取);to assert in the face of possible contradiction(易受反驳的声言);to claim to have(声称拥有);to assert to be rightfully one's own(宣称事实上属于自己的)。而名词claim的意思有:a demand for something due or believed to be due(要求某种应给予或认为应给予之物);a right to something(对某物的权利);an assertion open to challenge(易受反驳与争议的声言)[12]。

本合同条件1.1.3对Claim下的定义，既未提损失，亦未提"赔偿"。也就是说，《新华词典》对汉语"索赔"的解释，不适合本合同条件中Claim的定义，"索赔"不能反映claim原意。

Claim的上述定义还有一点值得注意，即，索要仅仅是单方提出的，对方不一定同意，也不一定得到公认。尽管援引了本条件的条文或提出了与本合同或实施本工程有关或由此产生的其他理由，最后发现站不住脚。

另一方面，读者不难从本条件中发现，雇主或承包人给对方造成损失，若原因或事实明确，都设有要求造成损失者给予受损方补偿的具体条文。雇主或承包人受损不能明确归因于对方，特别是因客观因素或第三方而受损的情况，都归入第20条[雇主与承包人索要]。雇主或承包人在第20条索要的理由均非对方直接给自己造成损失，因而谈不上"赔"。经过这样的比较，就会知道，以往汉译"索赔"曲解了claim原意，有误导之嫌。清华大学土木水利学院吴之明教授曾建议将Claim译成"索补"。仅就上文第1.1.3项中的定义，Claim的对象不仅有各种"救济"之物，还有非物的"权利"。本条件常提的"权利"很多是要求延长时间。"时间"不可逆，无法"补"，但可以延长。译者认为，"索要"在表达本合同条件中Claim的原意时，要比"索赔"或"索补"好。故此，以下将Claim译成"索要"。当然，见到claim（请注意，不是首字母大写的Claim）时，不一定一成不变地译成"索要"。最好视上下文，在不损害其原意的条件下，译成反映claim原意且易懂的汉文。】

1.1.8 "Contract Agreement" means the agreement entered into by both Parties in accordance with Sub-Clause 1.6 [Contract Agreement], including any annexed memoranda.

"合同协议书"指合同双方按照第1.6款[合同协议书]签署的协议书，包括附在其后的任何要点记录。

【译注：有人将memoranda与memorandum of understanding混淆。memoranda *n*. 1. an informal record;（非正式记录）*also*: a written reminder（书面提（醒）示）. 2. an informal written note of a transaction or proposed instrument.（交易或文书草案的非正式书面要点记录）[12]】

1.1.11 "Contractor" means the person(s) named as contractor in the Contract Agreement and the legal successors in title of such person(s).

"承包人"指合同协议书中称为承包人的人或多人，及其权利与义务的合法继承人。

【译注：title有如下意思：right (to sth., to do sth.)esp.(law) right to the possession of a position, property：权益、权利（与to连用，后接某事物，或与不定词连用，表示做某事）；（尤指法律上的）保有某地位、财产的权利，所有权。[13]上文中in title to such person(s)中的title与to连用，to后接的such person(s)并非"事物"。显然不能译成"保有对此（这些）人的所有权"。文献[6]、[7]和[8]将与"the legal successors in title to such person(s)"类似的短语"the legal successors in title to this person"译成了"财产所有权的合法继承人"[6][7][8]不知译文中"财产"二字从何而来。若将"in title to this person"译成"有此人地位"，就好理解了。文献[4]将"the legal successors in title to such person"译成"取得了此当事人资格的合法继承人"。[4]意思同译者的译文接近。

另外，承包人的"财产所有权的合法继承人"若将财产变卖，不再从事工程承包而是改行，从事其他事业，那么，本合同中承包人的义务和权力由谁继承呢？应当由"权利和义务的合法继承人"承担。】

1.1.26 "Dispute" means any situation where:
(a) one Party makes a claim against the other Party (which may be a Claim, as defined in these Conditions, or a matter to

be determined by the Employer's Representative under these Conditions, or otherwise);

(b) the other Party (if the Employer, under Sub-Clause 3.5.2 [Employer's Representative's determination] or otherwise) rejects the claim in whole or in part; and

(c) the first Party does not acquiesce (if the Contractor, by giving a NOD under Sub-Clause 3.5.5 [Dissatisfaction with Employer's Representative's determination] or otherwise), provided however that a failure by the other Party to oppose or respond to the claim, in whole or in part, may constitute a rejection if, in the circumstances, the DAAB or the arbitrator(s), as the case may be, deem it reasonable for it to do so.

"争议"指下列情况的任何一种：

（a）某方向另一方索要（可能是本条件定义的索要，或是须由雇主代表按本条件决定的某事，或者其他事项）；

（b）另一方（若是雇主，则按第3.5.2项[雇主代表决定]或以其他方式）拒绝该要求全部或部分；以及

（c）索方未容忍（若是承包人，则按第3.5.5项[对雇主代表决定不满]或以其他方式发出不满意通知），但是，若在上述情况下，避免/裁决争议小组或仲裁员，视情况而定认为另一方未反对或回应上述要求的全部或部分是合理的，则另一方这样做可以构成一种否决。

【译注：文献[3]至[5]，以及文献[13]都将dispute错译成了"争端"。《新华词典》争端：引起争执的事由。】

1.1.27 "Employer" means the person named as the employer in the Contract Agreement and the legal successors in title to this person.

"雇主"指合同协议书称为雇主的人，及其权利和义务的合法继承人。

【译注：有些人将Employer译成"业主"。《现代汉语词典》和《新华词典》业主：产业或企业的所有者。同承包人签订EPC合同者，往往不是建成后工程的所有者，仅仅是承包人的雇用者，所以，不宜将Employer译成"业主"。】

1.1.31 "Employer's Requirements" means the document entitled employer's requirements, as included in the Contract, and any additions and modifications to such document in accordance with the Contract. Such document describes the purpose(s) for which the Works are intended, and specifies Key Personnel (if any), the scope, and/or design and/or other performance, technical and evaluation criteria, for the Works.

"雇主要求说明书"指列入本合同以雇主要求说明书为名的文件，以及按照本合同对其所做的所有增补和修改。该文件说明了本工程预期的用途，具体说明了本工程可能会评价重要人员依据的准则，以及本工程范围，和/或设计和/或其他性能、技术与评价准则。

【译注：文献[8]将1999年第1版中1.1.1.3 "Employer's Requirements" means the document entitled employer's requirements, as included in the Contract, and any additions and modifications to such document in accordance with the Contract. Such document specifies the purpose, scope, and/or design and/or other technical criteria, for the Works.中的the purpose…for the Works.译成了"工程的目标"；将4.1中The Contractor shall design, execute and complete the Works in accordance with the Contract, and shall remedy any defects in the Works. When completed, the Works shall be fit for the purposes for which the Works are intended as defined in the Contract.中的When completed, the Works shall be fit for the purposes for which the Works are intended译成了"完成后，工程应能满足合同规定

的工程预期目的";5.1中(b) definitions of intended purposes of the Works or any parts thereof 中的intended purposes of the Works译成了"对工程或其任何部分的预期目的的说明",等。"工程"无生命,会有什么"目标"或"目的"呢?其实,上述几处的purpose(s)是"用途"或"用处"的意思。purpose *n.* the reason for which something exists or happens.(用途)[15]。国内外有些辞典未收录"用途"这一条目,例如文献[12]、[13]、[14]。文献[8]译者在这几处将purpose(s)译成"目标"或"目的",可能用的就是没有该条目的辞典。不过,即使所有辞典既有"目标"、"目的",又有"用途"条目,那么,在翻译时,应如何选用呢?那就要凭常识了,可谓"人有目的、目标,物有用途"。"工程的用途"是人实现目标,达到目的的工具或手段。】

1.1.33 "Exceptional Event" means an event or circumstance as defined in Sub-Clause 18.1 [Exceptional Events].

"异常事件"指第18.1款[异常事件]定义的某种事件或情况。

【译注:文献[16]将Exceptional Event译成"例外事件"。】

1.1.49 "Notice of Dissatisfaction" or "NOD" means the Notice one Party may give to the other Party if it is dissatisfied, either with an Employer's Representative's determination under Sub-Clause 3.5 [Agreement or Determination] or with a DAAB's decision under Sub-Clause 21.4 [Obtaining DAAB's Decision].

"不满意通知"(NOD)指合同一方若对雇主代表按第3.5款[商定或决定]决定之事,或避免/裁决争议小组按第21.4款[取得避免/裁决争议小组裁定]裁定之事不满,可能给另一方的通知。

【译注:文献[4]~[8]将DAAB's decision的decision译成"决定",并得到普遍认可。但是,若细想一下,还有可商榷之处。在汉语中,许多活动的参与者会做出多种"决定",不一定是在僵持不下的情况

下，由第三者做出。而DAAB's decision则是避免/裁决争议小组以第三者身份对于雇主和承包人争议之事做出的裁定。所以，为了区别第三者与参与者自己的决定，最好将其译成"裁定"。

文献[4]将determine和determination译成"确定"。在《现代汉语词典》和《新华词典》中，"确定"的意思是"明确而肯定"（形容词）和"明确地定下"（动词）。但是，determine和determination在FIDIC合同文件里是同agree和agreement相对而用的，Agreement or Determination（见下文第3.5款）。Agreement是承包人和雇主代表商议，取得一致时得到的结果，而Determination是承包人和雇主代表经过商议无法取得一致时，由雇主代表决定的结果，承包人一般是不同意的，常常引起争议，因而不能说"明确而肯定"。这样一来，将determine和determination译成"决定"要比"确定"好。】

1.1.52 "Performance Certificate" means the certificate issued by the Employer (or deemed to be issued) under Sub-Clause 11.9 [Performance Certificate].

"完工证书"指雇主按第11.9款[完工证书]发给或认为应发给的证书。

【译注：文献[5]将Performance Certificate译成了"履约证书"，而文献[6]、[7]和[8]将"Performance Certificate" means the certificate issued under Sub-Clause 11.9[*Performance Certificate*].译成了"'履约证书'系指根据第11.9款[履约证书]的规定颁发的证书。"译成"履约证书"是否恰当呢？请看第11.9款的译注。

请注意，2017年第2版将1999年第1版的certificate issued under Sub-Clause 11.9改成了certificate issued by the Employer (or deemed to be issued) under Sub-Clause 11.9。更容易看出，Performance Certificate不能译成"履约证书"。】

1.1.53 "Performance Damages" means the damages to be paid by the Contractor to the Employer for the failure to achieve the guaranteed performance of the Plant and/or the Works or any part of the Works (as the case may be), as set out in the Schedule of Performance Guarantees.

"性能赔偿金"指承包人因未实现性能保证书一览表所许诺的装备和/或工程或工程任何一部分（视情况而定）的性能而应付给雇主的赔偿金。

【译注：Performance Damages是2017年第2版的新术语。1999年第1版第12.4款用the relevant sum payable as non-performance damages for this failure表示这个意思，但是，non-performance易生误解。例如，文献[8]就将其理解为"未履约"。】

1.1.56 "Plant" means the apparatus, equipment, machinery and vehicles (including any components) whether on the Site or otherwise allocated to the Contract and intended to form or forming part of the Permanent Works.

"装备"指现场或别处分配用于本合同并准备或已形成永久工程一部分的用具、机具、机器与车辆（连同所有的配件）。

【译注：为了避免译成"设备"后，与utility、equipment、building services等混淆，将其译成"装备"。】

1.1.57 "Programme" means a detailed time programme prepared and submitted by the Contractor to which the Employer has given (or is deemed to have given) a Notice of No-objection under Sub-Clause 8.3 [Programme].

"当前实施计划"指承包人编制并提交，雇主按第8.3款[实施计划]发出（或认为已发出）不反对通知的详细时间安排计划。

【译注：这里疑原文有误，time可能属于衍文。下文第8.3款明确指出了Programme应当有的内容，远非time programme所能容。Programme与programme差别不仅仅在于首字母的大小写。若都译成"实施计划"，就会在汉文中模糊英文中Programme与programme

的差别，programme可能是过时的、修改过的或雇主不同意的。所以，将Programme译成"当前实施计划"，加"当前"二字以示区别。】

1.1.58 "Provisional Sum" means a sum (if any) which is specified in the Contract by the Employer as a provisional sum, for the execution of any part of the Works or for the supply of Plant, Materials or services under Sub-Clause 13.4 [Provisional Sums].

"暂定金额"指雇主按第13.4款[暂定金额]可能在本合同中具体指定为暂定金额，用于实施本工程任何一部分或用于供应装备、材料或公共设施补给的款项。

【译注：services做名词用，有如下意思：a facility supplying some public demands或system or arrangement that supplies public needs（满足公众需求或需要的系统或安排），例如，telephone services（电话设施），postal services（邮政设施）。再如，Building services,使建筑物与构筑物运转的建筑设备(services which make a building come to life)，有燃气、电与可再生供应；供暖与空调；水、排水与管线；自然与人工照明；扶梯与升降机；通风与制冷；通信线路；电话与信息技术网络；保卫与警报系统；火警与防火等（*The services include energy supply (gas, electricity and renewable sources); heating and air conditioning; water, drainage and plumbing; natural and artificial lighting; escalators and lifts; ventilation and refrigeration; communication lines; telephones and IT networks; security and alarm systems; fire detection and protection*）。可简单译成"建筑设备"，与Plant不同。

近些年来，不少地区用services表示service（服务）的多样含义，但易同表示"公共设施"的services混淆。在广泛用services表示"多样服务"含义的北美，为了区分service和services，常在表示"公共设施"的services前面加utility，即utility services。

文献[6]、[7]和[8]将services译成了"服务"。

services还有facility、system或arrangement使用者得到的满

足、便利、效用或补给之意。若在上文中将services仅译成"公共设施",则"供应公共设施"就难以理解。】

1.1.66 "Section" means a part of the Works specified in the Contract Data as a Section (if any).

"单项工程"指合同数据中可能具体说明为单项工程的本工程某一部分。

【译注:将Section译为"分项工程"或"单位工程"不妥。按照我国实行了几十年,已为人们所熟知的基本建设程序,一个建设项目,可由若干单项工程组成,各单项工程可以有若干单位工程,单位工程可分解为若干分部工程,分部工程分解成若干分项工程。分项工程在建设项目分解结构中最底层,与国际上工程量清单(Bill of Quantities)中的Item, Item of work对应。在绝大多数情况下,分项工程不能为雇主提供使用功能,雇主不会单独接收之。本条件第10.2款[接收部分工程]明确指出:"除非雇主要求说明书可能指明或双方可能商定,否则,雇主不得接收或使用本工程或单项工程的部分。"

这样一来,"Section"不能译为"分项工程"或"单位工程",可译成"单项工程"或"区段"。】

1.1.69 "Statement" means a statement submitted by the Contractor as part of an application for payment under Sub-Clause 14.2.1 [Advance Payment Guarantee] (if applicable), Sub-Clause 14.3 [Application for Interim Payment], Sub-Clause 14.10 [Statement at Completion] or Sub-Clause 14.11 [Final Statement].

"报表"指承包人在情况适用时按第14.2.1项[预付款归还保证书]、第14.3款[期中支付申请]、第14.10款[完工报表]或第14.11款[最终报表]申请付款时一块提交的某一报表。

【译注:译成汉语时,请注意Statement与statement的不同。

后者多数情况译成"声明"、"声明书",偶尔译成"财务报表"。】

1.1.73 "Tender" means the Contractor's signed offer for the Works, the JV Undertaking (if applicable) and all other documents which the Contractor submitted with the Tender (other than these Conditions, the Schedules and the Employer's Requirements, if so submitted), as included in the Contract.

"投标文件"指本合同中由承包人为本工程签署的要约、在情况适用时的联营体承诺书,以及承包人连同该要约一起提交的所有其他文件(即使提交了本条件、表单和雇主要求说明书,亦将其除外)。

【译注:submitted with the Tender在1999年第1版中是submitted therewith,很多人觉得submitted therewith难懂,翻译时难下笔。而submitted with the Tender则一目了然,好懂。

有些人将Tender译成了"投标函"、"投标书"、"标函"、"标书"[6][7]。汉语的"函"与"书"同义。若将"投标函"、"投标书"、"标函"、"标书"译成英语,应当是Letter of Tender或Letter of Bid。北美常用的bid与世界其他多数地区常用的tender同义。Letter of Tender是Tender的重要组成部分,但不是全部。FIDIC在1999年前出版的合同格式里,例如文献[4],没有Letter of Tender这个术语。Tender在合同文件中至少有三个含义。

当动词用时,是"愿意按言明的价钱做工、供货等(*tender means making an offer* (*to carry out work, supply goods, etc.*) *at a stated price*)",可译成"投标(报价)",这个含义大多数人不会弄错。

当名词用时,是"一方愿意供货或做某事而声明的价钱(*'tender' means a statement of the price at which one offers to supply goods or to do something*)[12]"。正是在这里,各人理解常常南辕北辙,有人理解为"价钱",即标价;有人理解为"声明书",即(投)标书。

文献[4]（下称"红皮书第4版"）第1.1款(b)(v)将*tender*定义为"承包商按照合同条文，为实施和完成工程并弥补其中所有缺陷，向雇主提出并为中标通知书所接受之价（the Contractor's priced offer to the Employer for the execution and completion of the Works and the remedying of any defects therein in accordance with the provisions of the contract, as accepted by the Letter of Acceptance）"。其中offer用作名词，意为"某人希望他人购买而提出的价钱(a price named by one proposing to buy)"

可是，红皮书第4版又将tender用作记载这个Contractor's priced offer的"声明书（*a statement of the price at which one offers to supply goods or to do something*）"的名称，即文献[6]中的Letter of Tender。

为了将人们从云里雾里解救出来，文献[6]添加了Letter of Tender，取代原来表示"声明书（statement of the Contractor's priced offer）"的Tender。

文献[6]还用Accepted Contract Amount取代红皮书第4版中的"承包人按照合同条文，为实施和完成工程并弥补其中所有缺陷，向雇主提出并为中标通知书所接受之价（the Contractor's priced offer to the Employer for the execution and completion of the Works and the remedying of any defects therein in accordance with the provisions of the contract, as accepted by the Letter of Acceptance）"。

在文献[9]的"投标人须知格式示例（*Example Form of Instructions to Tenderers*）"中用tender amount取代tender，表示投标人在Letter of Tender中要求得到的金额（the sum offered or the amount entered in the letter of tender）"。

既然Tender原有三个含义都已有了不致互相混淆的词语表达，那么，它还有何用呢？

文献[6]在1.1.1.8中将其定义为"由承包人随投标函一起提交并列入本合同的所有其他文件（the Letter of Tender and all other documents which the Contractor submitted with the Letter of

Tender, as included in the Contract)。"按照这个定义，将Tender译成"投标书"就不恰当了，因为"由承包人随投标函一起提交并列入本合同的所有其他文件"例如，投标函附录（见下文）、由投标人填写了价钱的工程量清单，动辄几十页，再加上施工组织设计（在FIDIC合同里称"实施计划"），工程由承包人设计的那一部分图纸，有些时候还有招标人请投标人提交的替代方案，投标保证，投标人提出的分包单位的资料，等等，数量不少，所有这些，不能以汉文"书（信、书籍）"冠名，而称之为"投标文件"更恰当。至于招标文件，英文就是tender documents或bidding documents。

至于Appendix to Tender，若因为将Tender译成"投标书"而将Appendix to Tender译成"投标书附录"，那就大错而特错。因为Appendix实际上是红皮书第4版中Tender的附录，即1999年第1版文献[6]中Letter of Tender的附录。实际上，在文献[6]的Letter of Tender中有如下明文：We acknowledge that the Appendix forms part of this Letter of Tender.（我们认可附录构成本投标函的一部分。）所以，若按1999年第1版编者的思路，Appendix to Tender的英文应当是Appendix to Letter of Tender。既然如此，是否应建议FIDIC合同委员会将其改过来呢？大可不必，只要将Tender译成"投标文件"而将Appendix to Tender译成"投标函附录"，就不会使人误以为Appendix是投标文件的"附录"。】

1.3 Notices and Other Communications通知与其他沟通讯息

Wherever these Conditions provide for the giving of a Notice (including a Notice of Dissatisfaction) or the issuing, providing, sending, submitting or transmitting of another type of communication (including acceptance, acknowledgement, advising, agreement, approval, certificate, Claim, consent, decision, determination, discharge, instruction, No-objection, record(s) of meeting, permission, proposal, record, reply, report, request, Review, Statement, statement, submission or any other similar type of communication), the Notice or other communication shall be in writing and:

本条件凡规定发出通知（包括不满意通知）或发出、准备、寄送、提交或传送另一种沟通讯息（包括认可、回辞、告知、认同、批准、证书、索要、应允、裁定、决定、解除、指示、不反对、会议记录、许可、建议、记录、回应、报告、要求、审核、报表、声明、提交物或所有其他类似沟通讯息）之处，通知与其他沟通讯息均应书写，并：

【译注1："通知"或"发出通知"，本条件大多数用give a Notice 或gaving of a Notice，而不是notify、inform、notice、advise等。原因在于：一，the Notice shall be in writing；二，"通知"对象拿到"通知"实物后，以后追究可有案可稽，而不能消失。

译注2："编写特别条文应注意之处"的Sub-Clause 1.2 Interpretation（解释），特别解释了"consent（应允）"的含义：In relation to the meaning of "consent" under sub-paragraph (g), it should be noted that this does not mean "approve" or "approval" which, under some legal jurisdictions, may be interpreted as accepting or acceptance that the requested matter is wholly satisfactory – following which the requesting party may no longer have any responsibility or liability for it.（对于g（段）中"应允"的含义，应注意，不是指在某些法律管辖区可以解释为请求之事令人完全满意，而后请求方对其不再负任何责任或负费的"接受"的"批准"。）

这个解释，把approve或者approval以及accepting或者acceptance与consent区别开来。

文献[3]至[5]都将consent译成了"同意"，这不但与从agree翻译过来的"同意"混淆，还曲解了consent的原意。】

1.9 Employer's Use of Contractor's Documents 雇主使用承包人文件

(b) entitle any person in proper possession of the relevant part of the Works to copy, use and communicate the Contractor's Documents and such other design documents for the purposes of completing, operating, maintaining, altering, adjusting, repairing and demolishing the Works;

（b）允许任何对本工程有关部分具有正当占有权之人为了完成、运行、维护、更改、调整、修复和拆除本工程复制、使用和传送承包人文件和上述其他设计文件。

【译注：for the purposes of意思是"为了……"，"对……来说"，"就……而言"。文献[8]将for the purposes of completing译成"为了完成……的目的"，没有必要。】

1.12 Compliance with Laws 遵守法律

(b) the Contractor shall give all notices, pay all taxes, duties and fees, and obtain all other permits, permissions, licences and/or approvals, as required by the Laws in relation to the execution of the Works. The Contractor shall indemnify and hold the Employer harmless against and from the consequences of any failure to do so unless the failure is caused by the Employer's failure to comply with Sub-Clause 2.2 [Assistance];

（b）承包人应发出所有通知，缴纳所有税金、关税和杂费，按法律对实施本工程的要求，取得所有其他许可证、许可、执照和/或批件。除非承包人办理上述各项时的延误或无果起因于雇主未遵守第2.2款[协助]，否则，承包人应保障雇主使其始终免受自己上述未果造成的后果。

【译注：fee *n.* a fixed charge; a sum paid for a service（手续费等，费）[12]fees（杂费）。charge *n.* the price demanded for something.（索价，要价，收取之费）charge *v.t.* to fix or ask as fee or payment（收取费用或款项）[12]】

(c) within the time(s) stated in the Employer's Requirements the Contractor shall provide such assistance and all documentation, as described in the Employer's Requirements or otherwise reasonably required by the Employer, so as to allow the Employer to obtain any permit, permission, licence or approval under sub-paragraph (a) above; and

（c）承包人应在雇主要求说明书指明的各时间内，按该文件的说明或雇主以其他方式合理提出的要求准备协助与所有文档，以便雇主能够取得上文（a）段中的任何许可证、许可、执照或批件；以及

【译注：the Contractor shall provide such assistance and all documentation中的provide是to make preparation to meet a need而非to supply something for sustenance or support的意思。不难理解，雇主在办理、领取许可证、许可、执照或批件的过程中请求承包人协助或给予文档，属于或有并非必有之事。只有需要时才会提出此等请求。若将provide译成"提供"，则会造成无论是否真的需要，也无论以何种方式，承包人都要协助或给予文档的误解。】

1.13 Joint and Several Liability 共同或多方负费

If the Contractor is a Joint Venture:

(a) the members of the JV shall be jointly and severally liable to the Employer for the performance of the Contractor's obligations under the Contract;

(b) the JV leader shall have authority to bind the Contractor and each member of the JV; and

(c) neither the members nor (if known) the scope and parts of the Works to be carried out by each member nor the legal status of the JV shall be altered without the prior consent of the

Employer (but such consent shall not relieve the altered JV from any liability under sub-paragraph (a) above).

若承包人属于联营体：

（a）该联营体成员在履行本合同承包人义务方面应共同和多方对雇主负费；

（b）联营体牵头人应有权约束承包人与联营体每一成员；并

（c）未经雇主事先应允（但此应允不应解除改变后联营体在上文（a）段中任何费负），既不得改变联营体成员，也不得改变应由各成员进行的本工程可能已经知道的范围和组成部分，亦不得改变联营体法律地位。

【译注：若将liability译成"责任"，将be liable for译成"（对）……负责"经常会同由responsibility译成的"责任"，由be responsible for译成的"（对）……负责"混淆，这样，就辜负了英文原文区分liability与responsibility，以及be liable for与be responsible for之间差别的苦心。Liability重点是支付"应付费用"、"应付代价"的责任，而responsibility重点是法律、合同、职责、道德责任。当然，若仅为了区分liability与responsibility，可将后者译成"义务"。然而，本合同条件中还有obligation，亦可译成"义务"。那么，又如何区分responsibility与obligation呢？若将liability、responsibility和obligation都译成"责任"，那就更糟了。另外，liability本来有"负债"之意。但"负债"在汉语中，常指已成的现状，很少人会理解成将来支付某种费用的责任。

笔者在权衡之后，决定如下翻译liability和be liable for：

若表示一种责任，则译成"负费"，其中"负"是动词，负担，承担；

若表示责任指向，则译成"费负"，其中"负"是名词，动词负担，承担的对象。具体视情况而定。】

1.14 Limitation of Liability 费负限度

Neither Party shall be liable to the other Party for loss of use of any Works, loss of profit, loss of any contract or for any indirect or consequential loss or damage which may be suffered by the other Party in connection with the Contract, other than under:

除了以下情况，任何一方不应因以任何方式使用本工程而造成的损失、利润损失、任何合同损失，或对方可能受到的同本合同有关的任何间接或引发的损失或损坏向其负费。

【译注：use of any Works不宜译成"使用任何工程"，本合同只有一个工程（当然，单项工程可有多个）。最好译成"以任何方式使用本工程"。】

1.15 Contract Termination 合同终止

Subject to any mandatory requirements under the governing law of the Contract, termination of the Contract under any Sub-Clause of these Conditions shall require no action of whatsoever kind by either Party other than as stated in the Sub-Clause.

除了本款指明的情况外，按本条件任何一款终止本合同时应符合本合同适用法的任何强制要求，并要求各方不起诉讼。

【译注1：Terminate *v.t.& i* to bring to an end; put an end to; come to an end: CLOSE[12] "终止合同"的"终止"，英文词除了terminate，还有determine、discharge和close等。

有些合同版本，例如FIDIC1987年以前的某些版本、爱尔兰现在的建筑师协会编的建筑合同（RIAI Building Contract），就用determine表示"终止合同"的"终止"。

determine *v.t.* to put or set an end to TERMINATE (~ an estate)[12]

例如，The contractor is entitled to seek payment for: The value of the work completed at the date of determination of the contract（承包人有权要求支付本合同终止之日已完成工作的价值）。但是，这些用determine表示"终止"的合同里还用determine表示本条件第3.5款[商定或决定]里的"决定"，使很多用户或其他读

者困惑难定。更有些合同，时而用determine，时而用terminate，有时还用discharge表示"终止"，混乱更甚。于是，从文献[4]开始，FIDIC就用determine表达本条件第3.5款[商定或决定]里的"决定"，用terminate表示"终止合同"的"终止"。译者写这段文字的用意在于提醒读者，在阅读其他国家或地区的其他组织编写的合同格式时，应注意determine的准确含义。

译注2：这一款是新添的，1999年第1版没有。添加的原因可能是，合同一方终止合同使对方受损，对方不能利用合同本身取得补偿，便诉诸仲裁或法院。于是，FIDIC在2017年第2版中添加了这一款，同时尽量完善有关终止合同的条文。】

However, if and to the extent that the Employer's failure was caused by any error or delay by the Contractor, including an error in, or delay in the submission of, any of the applicable Contractor's Documents, the Contractor shall not be entitled to such EOT and/or Cost Plus Profit.

但是，雇主若是因承包人任何错误或延误而未及时履约，包括任何影响雇主行动的承包人文件中错误或迟交，则承包人在上述后果范围内无权延长工期和/或索取费用加利润。

【译注：并非所有承包人文件中错误或迟交都会造成雇主未能及时给予承包人出入和占用之权，在这里，用applicable表示影响雇主上述行动的承包人文件中错误或迟交。

承包人的错误或延误造成雇主未及时履约，有程度或范围的差别，不能不管程度深浅或范围大小，就一概剥夺承包人延长工期，索取费用加利润的全部权利。to the extent that表示了承包人失去上述权利的多少要视承包人错误或延误造成雇主未及时履约的程度和范围而定。

文献[6]、[7]、[8]曾将1999年第1版与此对应的段落However, if and to the extent that the Employer's failure was caused by any error or delay by the Contractor, including an error in, or delay in the submission of, any of the Contractor's Documents, the Contractor shall not be entitled to such extension of time, Cost or profit. 译成"但是，如果出现雇主的违约是由于承包商的错误或延误，包括在任何承包商文件中的错误或提交延误造成的情况，承包商应无权得到此类延长期、费用或利润。"

该译文没有明确承包商"无权"的范围，读者会以为，只要承包商有错，不管大小与轻重，就一天也不能延长，一分钱也不能要，一棍子打死。造成误解的原因，就是没有将 to the extent that 的含义翻译出来。

另外，不难发现，第2版用 the applicable Contractor's Documents 取代了第1版中 the Contractor's Documents，缩小了因承包人文件有错误或迟交而取消承包人延长时间权利的范围。】

2.3　Employer's Personnel and Other Contractors 雇主人员与其他承包人

(a) co-operate with the Contractor's efforts under Sub-Clause 4.6 [Co-operation]; and

（a）按第4.6款[合作]与承包人各项努力配合；并

(b) comply with the same obligations which the Contractor is required to comply with under sub-paragraphs (a) to (e) of Sub-Clause 4.8 [Health and Safety Obligations] and under Sub-Clause 4.18 [Protection of the Environment].

（b）遵守第4.8款[健康与安全义务]（a）至（e）段及第4.18款[保护环境]要求承包人遵守的同样义务。

【译注：1999年第1版，(b)段的条文是(b) take actions similar to those which the Contractor is required to take under sub-paragraphs (a), (b)and(c) of Sub-Clause 4.8 [Safety Prbcedure] and under Sub-Clause 4.18 [Protection of the Environment]。也就是将 take actions 改成了 comply with obligations。"遵守义务"的要求比"采取行动"更高。】

The Contractor may require the Employer to remove (or

cause to be removed) any person of the Employer's Personnel or of the Employer's other contractors (if any) who is found, based on reasonable evidence, to have engaged in corrupt, fraudulent, collusive or coercive practice.

承包人可以要求雇主清除（或督促清除）雇主人员或雇主可能有的其他承包人中有合理证据表明从事腐败、欺诈、勾结或胁迫行为的任何人。

【译注：这段条文是2017年第2版新添加的。明文扩大了承包人权利。另外，世界上工程合同的雇主很大一部分是政府部门，而政府是易生腐败之地。添加这段条文，等于增加了反腐败的力量。】

2.4 Employer's Financial Arrangements 雇主资金准备

The Employer's arrangements for financing the Employer's obligations under the Contract shall be detailed in the Contract Data.

合同数据应详细说明雇主为履行本合同义务而筹集资金的准备。

【译注：文献[8]将arrangements译成了"安排"，有失妥当。请看汉语词典对"安排"的解释。《现代汉语词典》安排：有条理，分先后地处理（事务）；安置（人员）；措置；《新华词典》安排：根据情况有计划地对人或事作好安置处理。

再看英文词典对arrange和arrangement的解释。

arrange *v.t.* 1. To make preparations for:PLAN(arrange a reception for the visitor)（准备，等于PLAN（部署））2. To put into a proper order or into a correct or suitable sequence, relationship, or adjustment（使……处于恰当的秩序，或正确或适当的顺序、关系或调适态）[12]

arrangement *n.* the state of being arranged（安排后的状态）; act of arranging（准备或安排的行动）; a preliminary measure: PREPARATION（开始时的措施，等于准备）[12]。】

2.5 Site Data and Items of Reference 现场数据与参照物

The Employer shall have made available to the Contractor for information, before the Base Date, all relevant data in the Employer's possession on the topography of the Site and on sub-surface, hydrological, climatic and environmental conditions at the Site. The Employer shall promptly make available to the Contractor all such data which comes into the Employer's possession after the Base Date.

雇主应在基准日前准备好自己掌握的有关现场地形，以及地下和水文、气候及环境条件的所有数据，供承包人参考。同样，雇主也应立即为承包人使用而备齐基准日后得到的所有此类数据。

【译注：文献[8]将The Employer shall have made available to the Contractor for his information, prior to the Base Date, all relevant data in the Employer's possession on subsurface and hydrological conditions at the Site, including environmental aspects. The Employer shall similarly make available to the Contractor all such data which come into the Employer's possession after the Base Date译成了"雇主应在基准日期前，将其取得的现场地下、水文条件及环境方面的所有有关资料，提供给承包商。同样地，雇主在基准日期后得到的所有此类资料，也应提供给承包商。"

原文中的make available不宜翻译成"提供"，可译为"供……使用"。众所周知，"现场地下、水文条件及环境方面的所有有关资料"有些时候体积庞大、重量惊人。例如，长江三峡工程钻取的花岗岩芯，有的直径达2米。这些资料很多不能复制，有些非常宝贵，无法或不应交给承包人。只能请承包人到这些资料存放处察看、阅读等。将make available译作"提供"，易生误解。】

The Employer shall have no responsibility for the accuracy,

sufficiency or completeness of such data and/or items of reference, except as stated in Sub-Clause 5.1 [General Design Obligations].

除第5.1款[一般设计义务]指明外，雇主不负责上述数据和/或参照物准确、充分或完整。

【译注：accuracy、sufficiency和completeness都是表示"数据"和"参照物"状态、达到某种状态程度的词，而不表示性质。所以，不能译成"准确性"、"充分性"和"完整性"。

accuracy *n*. 1: freedom from mistake or error: CORRECTNESS 2: a: conformity to truth or to a standard or model: EXACTNESS b: degree of conformity of a measure to a standard or a true value.[12]

sufficiency *n*. 1: a sufficient means to meet one's needs: COMPETENCY; *also*: a modest but adequate scale of living 2: the quality or state of being sufficient: ADEQUACY.[12] sufficiency *n*.(usu. with indef. art.)~quantity（通常与不定冠词连用）足量；足够：*a sufficiency of fuel*.足够的燃料。[13]】

3.1 The Employer's Representative 雇主代表

The Employer's Representative shall be vested with, and (unless and until the Employer notifies the Contractor otherwise) shall be deemed to have, the full authority of the Employer under the Contract except in respect of Clause 15 [Termination by Employer].

除了第15条[雇主终止]，除非雇主另行通知承包人，应将雇主在本合同中的全部权限授予雇主代表并应认为其拥有之。

【译注：上面的条文比1999年第1版的条文Unless and until the Employer notifies the Contractor otherwise, the Employer's Representative shall be deemed to have the full authority of the Employer under Contract, except in respect of Clause

15[*Termination by Employer*]。多了"shall be vested with"。乍一看，拗口，难懂。其实，本来的逻辑是，如果雇主"不授予"，雇主代表怎么会"拥有"？"应认为雇主代表具有雇主在本合同中的全部权限"是告诉承包人，"请相信雇主代表"。但是，实践中，常有雇主代表授权不足的事情发生，使承包人受损。1999年第1版未提雇主的"授予"之责，只让承包人相信雇主代表，的确不够完善。】

3.5.5 Dissatisfaction with Employer's Representative's determination 对雇主代表决定不满

If the dissatisfied Party is dissatisfied with only part(s) of the Employer's Representative's determination:

(i) this part(s) shall be clearly identified in the NOD;

(ii) this part(s), and any other parts of the determination that are affected by such part(s) or rely on such part(s) for completeness, shall be deemed to be severable from the remainder of the determination; and

(iii) the remainder of the determination shall become final and binding on both Parties as if the NOD had not been given.

不满意一方不满意的若仅是雇主代表决定的部分：

（i）应在不满意通知中清楚地标识这些部分；

（ii）应当认为这些部分，以及该决定中受这些部分影响或为了完整而要依赖这些部分的其他部分是可以同该决定其余部分分割开来的；且

（iii）该决定其余部分应当如同未发出不满意通知一样属于最后的并对双方具有约束力。

【译注：近些年来，EPC合同及其管理越来越复杂，雇主代表在能力和业务水准上不能完全胜任，做出的决定往往不能使双方满意，所以，2017年第2版丰富了第3.5款的条文，为了行文方便，也为了合同当事人的方便，添加了NOD（不满意通知）这个新术语。】

4.1 Contractor's General Obligations 承包人一般义务

The Contractor shall provide the Plant (and spare parts, ifany) and Contractor's Documents specified in the Employer's Requirements, and all Contractor's Personnel, Goods, consumables and other things and services, whether of a temporary or permanent nature, required to fulfil the Contractor's obligations under the Contract.

承包人应准备雇主要求说明书具体说明的装备（以及各种可能有的备件）与承包人文件，以及所有承包人人员、货物、易耗品和为尽承包人本合同义务而必需的其他东西与临时或永久公共设施。

【译注：fulfil (or fulfill) v.t. *fulfill one's duties(a command, an obligation, sb's expectations of hopes)*（尽职（实行命令，尽义务，满足某人的期望））[13]

fulfil和perform在英汉词典里都有"履行"的释文。但是，两者含义不同。

FULFIL(FULFILL)implies a complete realization of ends or possibilities.（完全实现目的或可能性）而

PERFORM implies action that follows established pattern or procedure or fulfills agreed-upon requirements and often connotes special skill.（按照既定的模式或程序或实现商定的要求且常暗含特殊技能的行动）[12]

可以看出，fulfil强调结果，而perform多指过程。

为了将fulfil obligation和perform obligation区别开来，将fulfil obligation译成"尽义务"。《新华字典》：尽：完；全部用出；用力完成。】

The Contractor shall be responsible for the adequacy, stability and safety of all the Contractor's operations and activities, of all methods of construction and of all the Works.

承包人应负责承包人所有作业与活动、所有施工方法，以及整个工程的充分、稳定与安全。

【译注：文献[6]将这一段译成了"承包商应对所有现场作业、所有施工方法和全部工程的完备性、稳定性和安全性承担责任。"

adequacy、stability和safety指所有现场作业和所有施工方法，以及整个工程应当达到的状态，不是"性质"。不仅雇主担心承包人不能使现场作业、施工方法和工程达到这三种状态，而且承包人实际上也经常做不到。如果现场作业、施工方法和工程具备这三个"性质"，承包人也就无须再负责了。】

4.2 Performance Security 履约担保

The Contractor shall obtain (at the Contractor's cost) a Performance Security to secure the Contractor's proper performance of the Contract, in the amount and currencies stated in the Contract Data. If no amount is stated in the Contract Data, this Sub-Clause shall not apply.

承包人应自费按合同数据指出的金额和币种取得能够确保承包人履行本合同的履约担保，合同数据若未载明金额，则本款不适用。

【译注：文献[8]曾将The Contractor shall obtain (at his cost) a Performance Security for proper performance, in the amount and currencies stated in the Particular Conditions. If an amount is not stated in the Particular Conditions, this Sub-Clause shall not apply.译成了"承包商应对严格履约（自费）取得履约担保，保证金额和币种应符合专用条件中的规定。专用条件中没有提出保证金额的，本款应不适用。"

for proper performance 中的proper，意思是"恰当、适当"，没有"严格"的意思。另外，如何判断是否"恰当、适当"呢？只能依据合同。按照合同要求履行义务就是恰当，不存在严格不严格的问题。若译成"严格"，那么在实践中如何判断是否"严格"呢？

值得注意的是，2017年第2版，在proper performance之前添加了the Contractor's，并且将原来含糊不清的for改成了to secure。如此一来，这段话表达的意思，就比1999年第1版清楚多了。】

4.5.1 Objection to Nomination 反对指定

(c) the subcontract does not specify that, for the subcontracted work (including design, if any), the nominated Subcontractor shall:

(i) undertake to the Contractor such obligations and liabilities as will enable the Contractor to discharge the Contractor's corresponding obligations and liabilities under the Contract; and

（c）分包合同未具体说明该指定分包人应对分包的工作（包括可能的设计）：

（i）向承包人承诺使承包人能够履行其本合同中义务和费负的义务和费负，以及

【译注：discharge v.t. 履行，例如discharge the Contractor's corresponding obligations and liabilities（履行承包人相应的义务与费负）；使免除，使卸脱，例如discharge the Contractor from the corresponding obligations and liabilities。（解除承包人相应的义务与费负）。到底是译成"履行"，还是译成"解除、免除"，先看一看有无from出现。若有，译成"解除、免除"；若无，仔细看上下文。文献[4]曾将the nominated Subcontractor will undertake towards the Contractor such obligations and liabilities as will enable the Contractor to discharge his own obligations and liabilities towards the Employer under the terms of the Contract译成了"指定分包商应对承包商承担此类义务和责任，以使承包商可以按照此合同条款免除他自己对业主承担的义务和责任"。文献[4]译者可能未注意到这句话英文原文里并无from出现，因此不能译成"免除"。读者会在本条件下文第

18.6款见到discharge和from同时出现的情况。在本文件"Annex A EXAMPLE FORM OF PARENT COMPANY GUARANTEE 母公司保证书格式样本"中，读者会看到，虽然文中没有from，但是，根据上下文，也应将discharge译成"免除"。

另外，discharge和perform都可以译成"履行"，两者有什么区别？DISCHARGE implies execution and completion of appointed duties and tasks（执行并完成事先指定的职责和任务）；而PERFORM implies action that follows established pattern or procedure or fulfills agreed-upon requirements and often connotes special skill.（按照既定模式或程序或实现商定的要求且常暗含特殊技能的行动）[12]

可以看出，discharge既注重过程也注意结果的完善，而perform多指过程。

文献[6]将any failure by the Subcontractor to perform these obligations or to fulfil these liabilities译成了"分包商任何不能完成这些义务或履行这些责任"。

perform在这里没有"完成"的意思。文献[14]在perform条目下有"完成（事业等）"的释文，例句一个是perform a surgical operation on sb.给某人施行外科手术；一个是perform calculations with surprising speed.以惊人的速度完成演算。[14]其实，两个例句中的perform都可换成do，即do a surgical operation on sb.和do calculations with surprising speed.因而可分别译成"给某人做外科手术"或"给某人动外科手术"；"以惊人的速度进行计算"或"以惊人的速度演算"。那么，perform和do有什么区别呢？perform含义，已在第4.1款的译注说明过了，即PERFORM implies action that follows established pattern or procedure or fulfills agreed-upon requirements and often connotes special skill.（按照既定的模式或程序或实现商定的要求且常暗含特殊技能的行动）[12]也就是说，perform就是需要"特殊技能"的do。不难理解，"动手术"和"计算"都是需要"特殊技能"的。execution of the Works中的许多work也是需要"特殊技能"的，所以，第13.3.1项中有a description of the varied work performed or to be performed, 第14.6.2项中有perform any work的用法。

再看fulfil。

fulfil(or fulfill) *v.t.* to measure up to: SATISFY (例如, a federal institution orders your employer to withhold a certain portion of your wages to satisfy a debt you owe.偿还债务);satisfy *v.t.* to meet a financial obligation to (例如偿还债务)[12]

由此看来，在这里将fulfil these liabilities译成"偿还（清偿）债务"，而将any failure by the Subcontractor to perform these obligations or to fulfil these liabilities译成"该分包人以任何形式未履行或清偿（由本合同或与之有关产生的）义务或债务"是恰当的。

不过，用fulfil these liabilities表示"偿还（清偿）债务"的用法不多。】

4.8 Health and Safety Obligations 健康与安全义务

(f) provide fencing, lighting, safe access, guarding and watching of:

(i) the Works, until the Works are taken over under Clause 10 [Employer's Taking Over]; and (ii) any part of the Works where the Contractor is executing outstanding works or remedying any defects during the DNP; and

（f）设置下列各项的围栏、照明、安全出入口、保卫与观察点：（i）本工程，一直到按第10条[雇主接收]接收本工程为止；和（ii）本工程在缺陷通知期承包人实施未完工作或弥补任何缺陷的任何部分；以及

(g) provide any Temporary Works (including roadways, footways, guards and fences) which may be necessary, because of the execution of the Works, for the use and protection of the public and of owners and occupiers of adjacent land and property.

（g）因实施本工程而为保护公众和邻近土地和财产主人与占用人并供其使用设置可能需用的所有临时工程(包括道路、步行道、防护栏杆和围栏)。

【译注：文献[8] 将

(d) provide fencing, lighting, guarding and watching of the Works until completion and taking over under Clause 10 [*Employer's Taking Over*], and

(e) provide any Temporary Works (including roadways, footways, guards and fences) which may be necessary, because of the execution of the Works, for the use and protection of the public and of owners and occupiers of adjacent land.

译成了："

（d）在工程竣工和按照第10条[雇主接收]的规定移交前，提供围栏、照明、保卫和看守，（以及）

（e）因实施工程为公众和邻近土地的所有人、占有人使用和提供保护，提供可能需要的任何临时工程(包括道路、人行道、防护物和围栏)。"

provide本意是to make something available to[12]（准备或设置某物，在需要时就能用上，但现在不一定需要），与to supply or make available（*something wanted or needed*）[12]（因为现在就需要，提供了，就能满足需要）。所以，这里不能将provide译成"提供"。另外，which may be……, for the use……是any Temporary Works (including……)的定语。】

(iii) that are necessary to effect and maintain a healthy and safe working environment for all persons entitled to be on the Site and other places (if any) where the Works are being executed.

（ⅲ）为所有有权到现场与其他可能实施本工程的处所人员创造并维持健康且安全、行之有效的工作环境所需要。

【译注："创造环境"的"创造"，英文有多个词，例如，create、

set up、put up和effect等。但是，这里用了读者不熟悉的effect。effect *v.t.* to put into operation by a process successfully completed by means of inherent force in the agent capable of surmounting obstacles.（通过利用能够克服障碍的动因内在力量而成功完成的过程使之发挥作用）[12]。这就是说，创造这个环境，不仅仅是物质方面，还有人和制度因素，人的思想、习惯、行为和人际关系，承包人的企业与项目管理制度。】

4.9.2 Compliance Verification System 守约核实制度

The Contractor shall prepare and implement a Compliance Verification System to demonstrate that the design, Materials, Employer-Supplied Materials (if any), Plant, work and workmanship comply in all respects with the Contract.

承包人应制订并实施守约核实制度，表明设计、材料、可能有的雇主供料、装备、工作和工艺在所有方面符合本合同要求。

【译注：这段文字的意思很明白，该制度就是要核查承包人的设计、材料、可能有的雇主供料、装备、工作和工艺在所有方面是否符合本合同（合约）要求。但是，文献[16]将Compliance Verification System译成了"合规验证系统"。译文中的"规"不知指什么"规"，是法规吗，若是法规，应当交代来源。另外，Verification没有"验证"的意思。】

4.11 Sufficiency of the Contract Price 合同价足额

The Contractor shall be deemed to have satisfied himself/herself as to the correctness and sufficiency of the Contract Price stated in the Contract Agreement.

应当认为承包人已经使他/她相信合同协议书载明的合同价正确，足够使用。

【译注：correctness 和sufficiency（请见第2.5款译注）表示合同价与承包人为设计、实施和完成本工程并弥补所有缺陷需付出的

代价相比所处的状态，不是合同价的性质。文献[8]将correctness 和 sufficiency译成了"正确性和充分性"；将Sufficiency of Contract Price译成了"合同价的充分性"；而文献[4]将sufficiency of Tender 中的sufficiency译成了"完备性"。即使查阅国内编写的文献[14]，也不见sufficiency有带"性"的释文，因此，文献[4]和[6]的译法均非恰当。经验表明，若说"合同价"或标价（Tender）的性质，由于投标竞争激烈，投标人获标心切，再加上编制投标文件时间常常很紧迫，"合同价"或"标价"经常不充分，不完备，怎么会有一成不变的"充分性"或"完备性"呢？大多数合同格式设这一条，就是为了防止投标人中标并与雇主签订合同后，以合同价或标价不足为借口，向雇主追索工价。这第4.11款，就是如果事后发现合同价或标价果然不足时，雇主推脱补偿承包人责任的条文，也是提醒与敦促投标人（承包人）实事求是，认真、全面、周到地估算工程所费，不要心怀幻想，指望"低报价，后索赔"。】

Unless otherwise stated in the Contract, the Contract Price stated in the Contract Agreement shall be deemed to cover all the Contractor's obligations under the Contract and all things necessary for the proper execution of the Works in accordance with the Contract.

除非本合同另行指明，否则应当认为合同协议书载明的合同价对于履行承包人本合同中所有义务，以及按本合同恰当地实施本工程所必需的所有事情都足够使用

【译注：cover在这里的意思是"（钱）用于……而足够"，无"包括"、"涵盖"之意。若译成"包括"（如文献[4]和[6]），就无法让他人理解，在"合同价"与"义务"、"实施"、"完成"和"修补"这些不同质的事物之间何以能够互相"包括"？】

The Contractor shall take all necessary measures to prevent any road or bridge from being damaged by the Contractor's traffic or by the Contractor's Personnel. These measures shall include the proper use of appropriate vehicles (conforming to legal load and width limits (if any) and any other restrictions) and routes.

应认为承包人基准日已经对现场进出路线适用和可为己用感到满意。承包人应采取所有必要的措施，防止任何道路或桥梁受到承包人通行或承包人人员的损坏。这些措施应包括正确地使用适宜的车辆（符合可能有的法定载荷与宽度限制，以及所有的其他限制）和路线。

【译注：suitability和availability是指access routes to the Site满足承包人要求，以及是否可为其用的状态，不是access routes to the Site具有的性质。将其译成"适宜性"和"可用性"，有失恰当。】

4.19 Temporary Utilities 临时公用事业设备

The Contractor shall, except as stated below, be responsible for the provision of all temporary utilities, including electricity, gas, telecommunications, water and any other services the Contractor may require for the execution of the Works.

除下述情况外，承包人应负责准备自己为实施本工程可能需用的所有临时公用事业设备，包括电力、燃气、电信、水和任何其他公共设施补给。

The following provisions of this Sub-Clause shall only apply if, as stated in the Employer's Requirements, the Employer is to provide utilities for the Contractor's use. The Contractor shall be entitled to use, for the purposes of the Works, the utilities on the Site for which details and prices are given in the Employer's Requirements. The Contractor shall, at the Contractor's risk and cost, provide any apparatus necessary for the Contractor's use of these services and for measuring the quantities consumed. The apparatus provided for measuring quantities consumed shall be subject to the Employer's consent. The quantities consumed (if any) during each period of payment stated in the Contract Data (if not stated, each month) shall be measured by the Contractor, and the amount to be paid by the Contractor for such quantities (at the prices stated in the Employer's Requirements) shall be included in the relevant Statement.

只有雇主要求说明书指明由雇主准备公用事业设备供承包人使用时，本款下列条文才适用。

承包人为了本工程应有权使用雇主要求说明书中已有细节与价格的现场公用事业设备。承包人应自担风险和费用，准备为了使用这些设施，以及计量消耗量而需用的所有用具。为计量消耗量而准备的用具应得到雇主应允。合同数据中指明的每一付款期（若未指明，一个月）可能的消耗量应由承包人计量，而为该数量（按雇主要求说明书指明的价格）应由其支付给雇主的数额应列入有关报表。

【译注：文献[8]将provide译成"提供"，不恰当。"用具"是"承包人自己为了使用这些设施，以及计量消耗量而需用的"，并非"提供"他人。】

4.20　Progress Reports 进展报告

(h) comparisons of actual and planned progress, with details of any events or circumstances which may adversely affect the completion of the Works in accordance with the Programme and the Time for Completion, and the measures being (or to be) adopted to overcome delays.

（h）实际进展与计划进展的对比，连同可能损及按当前实施计划和工期完成本工程的任何事件或情况的详情，以及为克服延误正在(或即将)采取的各项措施。

【译注：1999年第1版的(h)条文是：comparisons of actual and planned progress, with details of any events or circumstances which may jeopardise the completion in accordance with the Contract, and the measures being (or to be) adopted to overcome delays. 2017年第2版将may jeopardise the completion in accordance with the Contract改成了may adversely affect the completion of the Works in accordance with the Programme and the Time for Completion。FIDIC这一改动的原因可能是，一般说来，the Programme不属于合同文件；所以，若以in accordance with the Contract当作对比的参考物，就太笼统，会使承包人茫然无措。而in accordance with the Programme and the Time for Completion却很具体。】

5.1 General Design Obligations 一般设计义务

The Contractor shall be deemed to have scrutinised, prior to the Base Date, the Employer's Requirements (including design criteria and calculations, if any). The Contractor shall carry out, and be responsible for, the design of the Works and for the accuracy of such Employer's Requirements (including design criteria and calculations), except as stated in this Sub-Clause below.

应当认为承包人已在基准日前细读了雇主要求说明书（包括可能列入其中的设计准则和计算书）。承包人应当设计本工程并为之负责，除了本款下文所说之外，还要负责雇主要求说明书（包括可能列入其中的设计准则和计算书）准确。

【译注：accuracy是表示"雇主要求说明书"一种状态，而非某种性质的词。所以，不能译成"准确性"。accuracy n. 1: freedom from mistake or error: CORRECTNESS 2: a: conformity to truth or to a standard or model: EXACTNESS b: degree of conformity of a measure to a standard or a true value.[12]】

Design shall be prepared by designers who:

图样与设计说明书应由下述设计人员编制：

(a) are engineers or other professionals, qualified, experienced and competent in the disciplines of the design for which they are responsible;

（a）合格、有经验且在其负责的设计专业胜任的工程师或其他专业人士；

【译注1：图样与设计说明书，我国古籍称为"图说"。例如，"嘉靖六年（1527年）六月丙午，总理河道侍郎章极等言，黄河济漕，固为国家之利，至于泛滥则为地方之患，今欲筑浚分杀，以免民患……乃为图说以闻。工部请从拯等议，上然之，命拯等刻期举工。"（《明世宗实录》卷77）

design这里是名词。design *n*. preliminary sketch or outline showing the main features of *something* to be executed.（表示准备实施的某物主要特征的初步图样或文字要点）[12]design *n*. drawing or outline from which *something* may be made.（设计图样，设计）outline *n*. statement of the chief facts, points, etc.（要点；大纲；纲要）[13]

译注2：对设计人员资质和行为提出要求的这段条文是2017年第2版新加的。1999年第1版对设计人员资质和行为没有要求。】

However, the Employer shall be responsible for the correctness of the following portions of the Employer's Requirements and of the following data and information provided by (or on behalf of) the Employer:

然而，雇主应负责雇主要求说明书如下部分和雇主（或以雇主名义）准备的下列数据和信息正确：

【译注：文献[8]将provided译成"提供"。雇主准备的数据或信息不一定都"提供"他人。请见第2.5款的译注1。另外，上文中的Any data or information received by the Contractor, from the Employer or otherwise也暗示了雇主掌握的数据或信息不一定都给了承包人。】

(d) portions, data and information which cannot be verified by the Contractor, except as otherwise stated in the Contract.

（d）除了本合同另外指出，承包人无法核实的部分、数据和信息。

【译注：文献[8]将 (a) portions, data and information which are stated in the Contract as being immutable or the responsibility of the Employer,译成了"(a) 在合同中规定的由雇主负责的，或不可变的部分、数据和资料，"immutable译成"不应变"比"不可变"好。若不加约束，雇主总想改变他原先准备的数据和资料。而设计人员却非常厌恶这样做，这些东西一变，设计就要变，常造成设计人员疲于奔命。承包人希望雇主承诺以后不要改变这些资料和数据，如果要改变，就要承担责任。另外，or在这里是"否则"而非"或者"的意思。】

5.2 Contractor's Documents 承包人文件

The Contractor's Documents shall comprise the documents:
承包人文件应由下列文件组成：

【译注：comprise *v.t.* to be made up of（由……组成）[12]include（包含；包括）; have as parts or members（以……为组成部分或成员；包括）; be composed of（由……组成）[13] 1. to consist of.（由……组成）2. *Informal*（非正式用法）a. to include or contain.（包含；包括）b. to form or constitute.（形成；构成）[15]尽管有些词典在comprise条目下有"包括；包含"的释文，但正如文献[5]指出的那样，属于非正式用法。

文献[8]将The Contractor's Documents shall comprise译成了"承包人文件应包括"值得商榷。《现代汉语词典》："包括"，包含（或列举各部分，或着重指出某一部分）。而"由下列文件组成"不但指全部，还暗指"下列文件"之间有某种使其组成的整体具备某种新功能、新用途的关系，然而，"包括"或"包含"却无指出其对象之间有某种关系的能力。

为了澄清读者的误解，防止include的滥用，2017年第2版在第1.2款（g）中特别解释了该词含义：

(h) "including", "include" and "includes" shall be

interpreted as not being limited to, or qualified by, the stated items that follow;

"包括"应理解为不限于下文指出的事项，或受其限制；】

(b) required to satisfy all permits, permissions, licences and other regulatory approvals which are the Contractor's responsibility under Sub-Clause 1.12 [Compliance with Laws]; and

（b）为符合办理所有许可证、许可、执照与其他政府批件的要求而需要，第1.12款[遵守法律]要求承包人负责者，以及

【译注：文献[8]将documents required to satisfy all regulatory approvals译成了"为满足所有规章要求报批的文件"。regulatory的意思是restricting according to rules or principles（按照规则或原则限制），是approvals的定语，而approvals是复数名词，不能译成动词"报批"。regulatory approvals应译成"受规章限制的批件"。若仍同文献[8]译者一样，将satisfy理解为"满足"，则satisfy all regulatory approvals应译成"满足所有受规章限制的批件"，但，很难理解。问题出在satisfy是"符合……要求"的意思，satisfy v.t. to conform to（符合（要求））[12] satisfy all regulatory approvals应当译成"符合办理所有受规章限制的批件的要求"。】

5.2.3 Construction 施工

(a) construction of such part shall not commence until a Notice of No-objection is given (or is deemed to have been given) by the Employer for all the Contractor's Documents which are relevant to its design and execution;

（a）只有雇主为有关该部分设计与实施的所有承包人文件已发出或认为已发出不反对通知时才能开始施工；

【译注：第1.2款[解释](j) "execution of the Works"已将design列为execution的一部分，这里又将design与execution并列，有失恰

当。可以改成its design and the other operations and activities of execution（指construction and completion of the Works and the remedying of any defects）。另见第7.1款及其译注。】

5.3 Contractor's Undertaking 承包人承诺

The Contractor undertakes that the design, the Contractor's Documents, the execution of the Works and the completed Works will be in accordance with:

承包人承诺，设计、承包人文件、本工程实施和完成后符合：

【译注：第1.2款[解释](j)"execution of the Works"已将design列为execution的一部分，这里又将design与execution of the Works并列，有失恰当。可以改成The Contractor undertakes that the Contractor's Documents, the execution of the Works, including the design, and the completed Works will be in accordance with:（承包人承诺，承包人文件、包括设计的本工程实施和完成后符合：）】

5.4 Technical Standards and Regulations 技术标准与法规

Regulation *n*. an authoritative rule dealing with details or procedures (safety regulations)（当局处理细节或程序的权威规则）; a rule or order issued by an executive authority or regulatory agency of a government and having the force of law.（行政当局或管理机构颁布的规则或命令）[12]a rule or order prescribed by authority, as to regulate conduct（权威机构为调整行为而制订的规则或命令）。[15]

【译注：文献[8]将Regulations译成"规定"，易与其他英文词译成的"规定"混淆。文献[7]将多个不同的英文词语译成了"规定"。这里只举若干例子。

Provisions，例如，第1条标题General Provisions（一般规定）；

Specify，例如，1.1.3.4 "Tests on Completion" means the tests

which are specified in the Contract or agreed by both Parties or…（1.1.3.4"竣工检验"系指在合同中规定或双方商定的，或……）；

Define，例如，1.1.4.3 "Final Statement" means the statement defined in Sub-Clause 14.11[Application for Final Payment].（1.1.4.3 "最终报表"系指第14.11款[最终付款的申请]规定的报表。）

provide for，例如，1.3 Communications（沟通）Wherever these Conditions provide for the giving or issuing of approvals, certificates, consents, determinations, notices, requests and discharges, these communications shall be:（本条件不论在何种场合规定给予或颁发批准、证书、应允、决定、通知和请求时，这些通信信息都应：）

state，例如，1.8 Care and Supply of Documents（照管与供应文件）…Unless otherwise stated in the Contract（……除非合同另有规定），等。

更重要的是，除了标准，本款条文未提笼统的"规定"，而是法规。《现代汉语词典》：法规：法律、条例、规则、章程的总称。《新华词典》：法规：法律效力低于宪法和法律的一种法的形式。在中国指国务院制定的行政法规和地方国家权力机关制定的地方性法规。一般用条例、规定、规则、办法等称谓。】

5.5 Training 培训

If the Employer's Requirements specify training which is to be carried out before taking-over, the Works shall not be considered to be completed for the purposes of taking-over under Sub-Clause 10.1 [Taking Over the Works and Sections] until this training has been completed in accordance with the Employer's Requirements.

如果雇主要求说明书具体指明须在接收前培训，则就第10.1款[接收工程与单项工程]的接收而言，只有按雇主要求说明书完成了这些培训时，才能认为本工程已经完成。

【译注：文献[8]将The Contractor shall carry out the training

of Employer's Personnel in the operation and maintenance of the Works to the extent specified in the Employer's Requirements. If the Contract specifies training which is to be carried out before taking-over, the Works shall not be considered to be completed for the purposes of taking-over under Sub-Clause 10.1 [*Taking Over of the Works and Sections*] until this training has been completed.译成了"承包商应根据雇主要求中规定的范围，对雇主人员进行工程操作和维修培训。如果合同规定了接收前要进行培训，在此项培训完成前，不应认为工程已根据第10.1款[工程和单位工程的接收]规定的接收要求竣工。"

operate或operation在这里的意思是to put or keep in operation，即"使……开始或继续运转"。如果operation and maintenance of 后面是a machine，译成"操作和维护该机器"，不会有问题，然而，operation and maintenance of 后面是the Works，而the Works可能是王府井的东方广场，也可能是三峡大坝，还可能是京沪高速铁路，这时再译成"操作和维护该工程"，听起来不顺耳。另外，shall not … until …的译法不太符合汉语习惯。

5.6 As-Built Records竣工记录

The Contractor shall prepare, and keep up-to-date, a complete set of "as-built" records of the execution of the Works, showing the exact as-built locations, sizes and details of the work as executed by the Contractor.

承包人应当编制并随时更新一套完整的本工程实施"竣工记录"，按承包人实施后原状显示工作的准确竣工部位、大小和细节。

【译注：文献[8] 将The Contractor shall prepare, and keep up-to-date, a complete set of "as-built" records of the execution of the Works, showing the exact as-built locations, sizes and details of the work as executed. 译成了"承包商应当编制并随时更新一套完整的工程实施"竣工"记录，如实记载竣工

的准确位置、尺寸和已实施工作的详细说明。"

这里首先的问题是"如实记载竣工的准确位置、尺寸"含义不清，是什么东西的"竣工的准确位置、尺寸"呢？第二个问题，of the work as executed不仅仅是details的后置定语，而是the exact as-built locations, sizes and details三者的后置定语。第三个问题是as executed应当如何翻译？若像文献[8]那样，译成"已实施工作"，就会有疑问，应当在何时记载"已实施工作"呢？应当是在实施后，趁着还没有发生任何变化时记载。所以，翻译的时候，应将这层意思翻译出来。】

The format, referencing system, system of electronic storage and other relevant details of the as-built records shall be as stated in the Employer's Requirements (if not stated, as acceptable to the Employer). These records shall be kept on the Site and shall be used exclusively for the purposes of this Sub-Clause.

雇主要求说明书应指明竣工记录的格式、索引方法、电子存储方法和其他有关细节。(若未指明，应得到雇主认可)。现场应始终保存这些记录并仅为本款所用。

【译注：system n. any formulated mothod of plan（任何有条理的方法或计划），有"方法"的意思。referencing system是图样、各种文献的"索引方法"，不是"基准体系"或"参考系统"。】

The Contractor shall submit to the Employer under Sub-Clause 5.2.2 [Review by Employer]:

承包人应按第5.2.2项[雇主审核]向雇主提交：

(a) the as-built records for the Works or Section (as the case may be) before the commencement of the Tests on Completion; and

（a）本工程或单项工程（视情况而定）开始完工检验前的竣工记录，以及

(b) updated as-built records to the extent that any work is executed by the Contractor:

（b）由承包人如下时间实施的任何工作的更新竣工记录：

【译注：to the extent that是"在……范围内"的意思，限制了"更新竣工记录"的范围，不是"由承包人实施的任何工作"的"更新竣工记录"，而是"由承包人如下时间实施的任何工作"的"更新竣工记录"。】

6.1 Engagement of Staff and Labour 聘用员工

Except as otherwise stated in the Employer's Requirements, the Contractor shall make arrangements for the engagement of all Contractor's Personnel, and for their payment, accommodation, feeding, transport and welfare.

除雇主要求说明书另有说明外，承包人应为所有承包人人员的聘用、薪酬、住宿、膳食、交通和福利做好准备。

【译注：make arrangements最好译成"做准备"，可见第2.4款的译注。在一些英汉辞典里，将engage和employ都译成"雇用"。但两者含义不同。

engage *v.t.* (engagement *n.*) to provide occupation for: INVOLVE(engage him in a new project)（为……设置职位，等于involve（使……参与；使……从事（聘他参加新项目）））[12] to obtain the right to employ.（取得雇用的权利，聘用）[13]to secure for employment, use, aid, etc.（为雇用、使用、借助，等而取得，聘用）[15]

employ *v.t.* to give work to, usually for payment（将工作给予，一般都要付酬，雇用）[13]

engage是employ的前提，employ是engage的目的。只有聘用了（having engaged）员工（having engaged），才能使（雇）用（employ）员工。

文献[8]将engage(ment)译成了"雇用"，容易与从employ(ment)译成的"雇用"混淆。】

6.2 Rates of Wages and Conditions of Labour 工资单价和劳动条件

【译注：rates of wages这里指daily, weekly, monthly or quarterly wages，可译成"工资单价"。】

The Contractor shall not recruit, or attempt to recruit, staff and labour from amongst the Employer's Personnel.

承包人不得从雇主人员中补充自己的员工，也不应有此种企图。

【译注：文献[8]将recruit译成"招收"。《现代汉语词典》招收：用考试或其他方式接收（学员、学徒等）recruit v.t. to engage or hire (new employees, members,etc.)（聘用或（短期或不定期）雇用（雇员、成员，等））[15]】

6.8 Contractor's Superintendence 承包人统筹督导

From the Commencement Date until the issue of the Performance Certificate, the Contractor shall provide all necessary superintendence to plan, arrange, direct, manage, inspect, test and monitor the execution of the Works.

从开工日直到颁发完工证书，承包人都应具备统筹督导规划、准备、指导、控制、检查、检验和监视本工程实施的所有必要职能。

【译注：manage v.t. to handle or control in action or use（在行动或使用中处理或控制）[15]不能译成"管理"。管理学中，规划、准备、指导、控制、检查、检验和监视都属于"管理"，在行文中，不能将部分（规划、准备、指导、控制、检查、检验和监视）与整体（管理）并列。文献[6]、[7]和[8]将Contractor's Superintendence译成了"承包商的监督"。superintendence的原意是"*act or functions of superintending or directing*"[12]（监督或指导的行动或职能），在本款中，意思是"规划、准备、指导、控制、检查、检验和监视的行动或职能"。provide的意思是"设置，配备"，"具备"是"设置"的结果。

另外，2017年第2版将1999年第1版的the Contractor shall provide all necessary superintendence to plan, arrange, direct, manage, inspect and test the work.改成了the Contractor shall provide all necessary superintendence to plan, arrange, direct, manage, inspect, test and monitor the execution of the Works.也就是说，为Superintendence添加了monitor职能。而plan、arrange、direct、manage、inspect、test和monitor的对象也扩大了，不再仅仅是work，而是execution of the Works。

如此看来，将Superintendence译成"监督"，过于简单，易引起误解。

笔者将Superintendence译成"统筹督导"。在汉语里，"统"是"统领"的意思；"筹"是"规划、筹划、安排"的意思；"督"是"monitor, inspect, test"；"导"是"direct"。译者现在还没有办法将plan, arrange, direct, manage, inspect, test and monitor译成一个简单的汉语词。】

7.1 Manner of Execution 工作态度与行为方式

The Contractor shall carry out the manufacture, supply, installation, testing and commissioning and/or repair of Plant, the production, manufacture, supply and testing of Materials, and all other operations and activities during the execution of the Works:

在实施本工程期间，承包人应以如下态度与方式制造、供应、安装、检验与试用和/或修理装备；生产、加工、供应和检验材料，以及进行所有其他作业和活动：

【译注：文献[6]、[7]和[8]将Manner of Execution译成了"实施方法"。在本合同条件1999年第1版，以及2017年第2版，"实施方法"是第8.3款[实施计划]的内容，不应再设第7.1款另外讨论"实施方法"。Manner固然有"方法"之意，但也有"方式"、"态度"、"举止"、"风度"和"习惯"之意。众所周知，材料、工艺、工作、工程的质量，在很大程度上与工程实施者的"方式""态度"和"习

惯"有关，第7.1款就是针对这一问题而设的。所以，将Manner of Execution译成"工作态度与行为方式"要比"实施方法"更能反映原意。

1999年第1版第7.1款第一句话是：

The Contractor shall carry out the manufacture of Plant, the production and manufacture of Materials, and all other execution of the Works:（承包商应在制造装备、生产和加工材料，以及在工程所有其他作业中：）其中，all other execution具体含义不清。现在，2017年第2版阐明了all other execution的具体含义：即manufacture, supply, installation, testing and commissioning and/or repair of Plant, the production, manufacture, supply and testing of Materials, and all other operations and activities during the execution of the Works。确实比第1版好理解。】

7.4 Testing by the Contractor 承包人检验

The Contractor shall provide all apparatus, assistance, documents and other information, temporary supplies of electricity and water, equipment, fuel, consumables, instruments, labour, materials, and suitably qualified, experienced and competent staff, as are necessary to carry out the specified tests efficiently and properly.

承包人应准备使本合同具体说明的各项检验恰当而又时费消耗少所必需的所有器具、协助、文件和其他资料、临时水电供应、机具、燃料、易耗品、仪器、劳动力、材料，以及具有适当能力、经验和胜任的工作人员。

【译注：文献[8]将provide译成"提供"，不恰当，理由见第4.19款[临时公用事业设备]译注。

efficiently不同于effectively。

efficient *adj.* productive without waste（具有无枉费产出能力的）[12] performing or functioning in the least wasteful manner（以

最少耗费发挥作用的)[15]

effective *adj.* producing a decided, decisive, or a desired effect(产生某决定或希望效果的)[12] producing an expected effect(产生某预期效果的)[15]】

The Contractor shall promptly forward to the Employer duly certified reports of the tests. When the specified tests have been passed, the Employer shall endorse the Contractor's test certificate, or issue a test certificate to the Contractor, to that effect. If the Employer has not attended the tests, the Employer shall be deemed to have accepted the readings as accurate.

承包人应立即向雇主转交得到正当证明的检验报告。当通过合同具体说明的检验时,雇主应签字认可承包人的检验证书,或发给承包人说明已经通过合同具体说明的检验的检验证书。如果雇主未参加检验,应认为雇主已经承认记录检验结果的数据是准确的。

【译注1:文献[6]至[8]将The Contractor shall promptly forward to the Engineer duly certified reports of the tests.译成了"承包商应立即向工程师提交充分证实的试验报告。"这里,forward是"转交"的意思,与submit不同。duly certified reports of the tests一般是独立的检验机构开具后交给承包人,承包人再转交雇主(工程师)。

译注2:文献[8]将issue a test certificate to the Contractor, to that effect.译成了"向承包商颁发等效的证书。"to that effect是"用那个意思,带有那个意思",而"那个意思"就是在to that effect前面说过的"雇主认可承包人的试验证书",to that effect与"等效"毫不相干。】

8.1　Commencement of Works　开工

The Contractor shall commence the execution of the Works on, or as soon as is reasonably practicable after the

Commencement Date, and shall then proceed with the Works with due expedition and without delay.

承包人应在开工日或开工日后尽早开始实施本工程,然后将其从速、不耽搁地继续下去。

【译注：文献[8]将The Contractor shall commence the design and execution of the Works as soon as is reasonably practicable after the Commencement Date, and shall then proceed with the Works with due expedition and without delay译成了"承包商应在开工日期后,在合理可能情况下尽早开始工程的设计和施工,随后应以正当速度,不拖延地进行工程"。没有将proceed with的"继续进行"的意思翻译出来。很多雇主不但担心承包人迟迟不开工,而且担心开工以后断断续续,三天打鱼两天晒网,所以,上面的条文就是要求承包人尽早开工,开了工就不要停顿。】

8.5 Extension of Time for Completion 延长工期

The Contractor shall be entitled subject to Sub-Clause 20.2 [Claims For Payment and/or EOT] to Extension of Time if and to the extent that completion for the purposes of Sub-Clause 10.1 [Taking Over the Works and Sections] is or will be delayed by any of the following causes:

下列任何一个原因若就第10.1款[接收工程与单项工程]而言已经或将延误完工时间,则承包人在满足第20.2款[索要款项与/或延长工期]时,有权在此延误范围内延长工期:

【译注：文献[6]、[7]和[8]都将The Contractor shall be entitled subject to Sub-Clause 20.1[*Contractor's Claims*] to an extension of the Time for Completion if and to the extent that completion for the purposes of Sub-Clause 10.1 [*Taking Over of the Works and Sections*] is or will be delayed by any of the following causes:译成了"如果由于下列任何原因,致使达到按照第10.1款[工程和分项工程的接收]要求的竣工受到或将受到延误的程度,承包商有权按照第20.1款[承包

商的索赔]的规定提出延长竣工时间："。

to the extent that在这里的意思是"在这样的范围内（条件下）"，限定The Contractor shall be entitled……to an extension of……的范围。不是"达到这样的程度以至"的意思。

除了"下列任何原因（any of the following causes）"延误了时间，承包人应当负责的原因也会延误时间。比如说，下列任何原因延误了一周，承包人自己延误了5天，承包人就只有权要求延长一周，无权要求延长12天。所以，译文要反映这个意思。而文献[6]、[7]和[8]不但都没有这层意思，反而是"'任何原因致使达到按照第10.1款要求的竣工受到或将受到延误的程度'时，承包商就有了延长时间的权利"的意思，这不是同英文原意正相反吗？】

(a) the work on Plant, or delivery of Plant and/or Materials, has been suspended for more than 28 days and

（a）装备上（或制造装备）的工作，或装备和/或材料的送交已暂停28天以上，以及

【译注：文献[8]将the work on Plant译成了"生产设备的生产"。这里的work，无论是从工序、过程，还是从耗费的时间来看，只是"生产设备"生产过程的一部分；另一方面，在多数情况下，"暂定"不会从开工开始，一直到该"生产设备"完成，所以，将其译成"生产"有失恰当。另外，work有汉语中"活儿"的意思，the work on Plant就是"装备上的活儿"，有备料、下料、制模、铸造、车、钳、铣、刨、铆、焊、磨、镀、油漆、组装、调试，等。】

9 Tests on Completion 完工检验

【译注：文献[6]至[8]，以及[16]都将test译成了"试验"。

Test *n.* examination or trial (of sth.) to find its quality, value, composition, etc.[13]《新华词典》试验：为考察某事物的效果或性能而先在实验室或小范围内试作。《新华词典》检验：检查、验证。《现代汉语词典》验证：证验。

从第9条条文可明显看出，将test译成了"检验"要比"试验"

恰当。】

9.1 Contractor's Obligations 承包人义务

The Contractor may apply for a Taking-Over Certificate by giving a Notice to the Employer not more than 14 days before the Works will, in the Contractor's opinion, be complete and ready for taking over.

承包人可在他认为本工程即将完成并做好接收准备前最多14天，向雇主发出通知，申请接收证书。

【译注：文献[8]将1999年第1版中的not earlier than 14 days before the Works will, in the Contractor's opinion, be complete and ready for taking over.译成了"承包商可在他认为工程将竣工并做好接收准备的日期前不少于14天"。这里可以画图，看看这种译法是否准确：

承包商认为工程将竣工并做好接收准备的日期前20天，即6月1日	承包商认为工程将竣工并做好接收准备的日期前14天，即6月6日	承包商认为工程将竣工并做好接收准备的日期前10天，即6月10日	承包商认为工程将竣工并做好接收准备的日期，比如说6月20日

请问，允许承包人哪一天通知雇主，6月1日，6月6日，还是6月10日？6月19日行不行？

从英文来看，6月6日和6月10日都是允许的，6月19日也行。不允许6月1日。

读者会问，通知不是越早越好吗？其实，承包人申请接收证书的心情一般都是急迫的，希望早点通知雇主。但是，雇主的心情不同，不希望承包人申请太早，原因种种，例如，太早了容易遗忘。如此看来，文献[8]的译法不对。但是情有可原，因为很多人不明白not earlier than 14 days before的缘故。所以，2017年第2版将其改成了not more than 14 days before，即使不明白其中缘故，但是字面是明确的："……前不多于14天"。】

11.3 Extension of Defects Notification Period 缺陷通知期延长

(a) if and to the extent that the Works, Section (or the part of the Works) or a major item of Plant (as the case may be, and after taking over) cannot be used for the intended purpose(s) by reason of a defect or damage which is attributable to any of the matters under sub-paragraphs (a) to (d) of Sub-Clause 11.2 [Cost of Remedying Defects]; and

（a）本工程、单项工程（或部分工程）或某件主要装备（视情况而定，且在接收后）因某项可归咎于第11.2款[修补缺陷费用]（a）至（d）段的任何一事的缺陷或损坏不能用于各原定用途；且

【译注：文献[8]将1999年第1版的The Employer shall be entitled subject to Sub-Clause 2.5 [*Employer's Claims*] to an extension of the Defects Notification Period for the Works or a Section if and to the extent that the Works, Section or a major item of Plant (as the case may be, and after taking over) cannot be used for the purposes for which they are intended by reason of a defect or damage.译成了："如果因为某项缺陷或损害达到使工程、单位工程或某项主要生产设备（视情况而定，并在接收以后）不能按原定目的使用的程度，雇主应有权根据第2.5款[雇主的索赔]的规定对工程或某一单位工程的缺陷通知期限提出一个延长期。"

这样翻译，不但没有反映"雇主延限的权利受工程等不能按原定目的使用的程度的限制"的意思，反而是"只要缺陷或损害使工程等不能按原定目的使用，雇主就有了延期权利"的意思。】

11.4 Failure to Remedy Defects 未能修补缺陷

(i) payment of Performance Damages by the Contractor in full satisfaction of this failure; or （i）承包人支付足以补偿其未修补缺陷或损坏后果的性能损害赔偿金；或

【译注：satisfaction在这里是"补偿"的意思。satisfy v.t. to make reparation to (an injured party)（补偿受损者）[12] satisfy v.t. to make reparation to or for（补偿）[15]】

11.5 Remedying of Defective Work off Site 现场外修补有缺陷工作

The Contractor shall also provide any further details that the Employer may reasonably require.

承包人还应按雇主可能有的合理要求准备任何进一步细节。

【译注：工程用的装备和/或材料，很多量多、体巨、构造复杂，承包人无法将其从装备或材料所在地点"提供"给雇主，可能需要雇主亲到装备或材料所在地点查看，这样，承包人就要准备好细节，让雇主查看。】

11.9 Performance Certificate 完工证书

Only the Performance Certificate shall be deemed to constitute acceptance of the Works.

应认为，只有完工证书才构成对本工程的认可。

【译注：最后"应认为，只有完工证书才构成对本工程的认可。"这句话的言外之意是"不能认为缺陷通知期开始前颁发的'接收证书'构成对本工程的认可"，"完工证书（Performance Certificate）才是本工程已经完成的证明"。如此说来，将Performance Certificate译成"完工证明"或"完工证书"就顺理成章。若译成"履约证书"，给人的感觉是"承包商履约的证书"，那是不是"资质证书"、"营业执照"，等等呢？

顺便提及，文献[4]中没有Performance Certificate这一术语，而是用Defects Liability Certificate（（解除）缺陷责任证书）"构成对本工程的认可"，即Only the Defects Liability Certificate, referred to in Clause 62, shall be deemed to constitute approval of the Works.（只有第62条提到的（解除）缺陷责任证书才能认为是构成对本工程的认可。）"（解除）缺陷责任证书"这一术语很难理解，所以FIDIC在文献[5]-[8]中将其改成Performance Certificate。】

11.10 Unfulfilled Obligations 未尽义务。

【译注：文献[8]将Unfulfilled Obligations译成"未履行的义务"，不恰当。英文中unperformed与unfulfilled不同。Unperformed是"根本未履行"，而unfulfilled是"履行了，但不彻底，未尽"。】

13.1 Right to Vary 变更权利

Variations may be initiated by the Employer under Sub-Clause 13.3 [Variation Procedure] at any time before the issue of the Taking-Over Certificate for the Works.

颁发本工程接收证书之前任何时候，雇主都可按第13.3款[变更程序]发起变更。

【译注：initiate *v.t.* to cause or facilitate the beginning of: set going(initiate a program of reform)（促使……开始；启动（发起一改革计划））[12]begin; set (a scheme, etc.)working（开始；着手）[13]】

Other than as stated under Sub-Clause 11.4 [Failure to Remedy Defects], a Variation shall not comprise the omission of any work which is to be carried out by the Employer or by others unless otherwise agreed by the Parties.

与第11.4款[未修补缺陷]指明的不同，除非双方另外商定，变更不能由省略任何要由雇主或他人进行的工作组成。（省略任何要由雇主或他人进行的工作不能构成变更）

【译注：文献[8]将1999年第1版中的a Variation shall not comprise the omission of any work which is to be carried out by others译成了"变更不应包括准备交他人进行的任何工作的删减"。将

comprise译成"包括",容易同从include译出的"包括"混淆。第5.2款的译注已经解释了"包括"或"包含"与"由……组成"含义的区别。】

13.3.1 Variation by Instruction 按指示变更

(a) a description of the varied work performed or to be performed, including details of the resources and methods adopted or to be adopted by the Contractor;

（a）对已做或应做的变更工作的说明,包括承包人已或将使用的资源与采用的方法;

(b) a programme for its execution and the Contractor's proposal for any necessary modifications (if any) to the Programme according to Sub-Clause 8.3 [Programme] and to the Time for Completion; and

（b）实施该变更的计划与承包人按第8.3款[实施计划]与工期对当前实施计划可能做的任何必要修改的提议;以及

【译注:perform *v.t.* carry out, do.[12] 进行,做。do (a piece of work, sth. one is ordered to do, sth. one has promised to do) 做（一件工作,一人奉命做的某事,一人答应做的某事）,履行,执行。[13]

according to和in accordance with都译成汉语"按照"、"依照"、"依据"、"与……一致"、"与……相应"。但是,according to与in accordance with 的不同之处在于according to还有"随……的不同（而不同）"的意思。所有,在用英文表达汉语中"按照"、"依照"或"依据"时,要注意according to与in accordance with 的区别。多数情况下,用in accordance with,而不用according to。

另一方面,according to在本文件中只用了三次。

第一次是在"1.1.46 'month' is a calendar month (according to the Gregorian calendar). '月'指根据格里历计数的太阳历月。"

第二次,就是本款,"any necessary modifications (if any)

to the Programme according to Sub-Clause 8.3[Programme]按第8.3款[实施计划]对当前实施计划任何必要的修改"。容易理解,按第8.3款[实施计划]"对当前实施计划任何必要的修改"确实是"随着(工程实际进展与实施计划本身)的不同而不同",所以"按第8.3款[实施计划]"用的是according to Sub-Clause 8.3,而不是in accordance with Sub-Clause 8.3。

第三次,在本文件"Annex A EXAMPLE FORM OF PARENT COMPANY GUARANTEE 母公司保证书格式样本"中,"the Contractor's compliance with all its terms and conditions according to their true intent and meaning.承包人按所有条款与条件的本意与含义遵守之"。也就是说,承包人是否遵守所有条款与条件,要看这些条款与条件的本意与含义。也确实是说,"是否遵守,随着(条款与条件的本意与含义)的不同而不同"。

至于第一次,"月"指根据格里历计数的太阳历月,也就是说,一年十二个月,各月是不同的,有28天、30天与31天的区别,"随着(月)的不同而不同"。】

- the Contractor has incurred or will incur cost which, if the work had not been omitted, would have been deemed to be covered by a sum forming part of the Contract Price stated in the Contract Agreement; and

- 承包人已经或将要开销的费用,若不省略该工作,认为可从构成合同协议书所载合同价一部分的某数额中支用;以及

【译注:cover的意思和用法,请见第4.11款译注。】

13.4 Provisional Sums 暂定金额

(ii) a sum for overhead charges and profit, calculated as a percentage of these actual amounts by applying the relevant percentage rate (if any) stated in the applicable Schedule. If there is no such rate, the percentage rate stated in the Contract Data shall be applied.

（ⅱ）利用相应表单中可能指出的百分比乘上上述实际数额而算得，用作管理费与利润的份额。若无这种百分比，应利用合同数据中指明的百分比。

【译注：文献[6]、[7]和[8]都将1999年第1版中的a sum for overhead charges and profit, calculated as a percentage of these actual amounts by applying the relevant percentage rate (if any) stated in译成了"以……规定的有关百分率（如果有）计算的，这些实际金额的一个百分比，作为管理费和利润的金额"。按常识，"一个百分比"无法"作为……金额"。问题出在a percentage of these actual amounts 不能译成"实际金额的一个百分比"，可以译成"在实际金额中占的份额"，等。

percentage rate *n.*（百分比；百分率）percentage *n.* the result obtained by multiplying a number by a percent（将某数乘上一个百分比值后得到的结果，百分数）[12] percentage *n.* a proportion in general（在总体中占的份额）。[15]】

13.7 Adjustments for Changes in Cost 因费用改变的调整

To the extent that full compensation for any rise or fall in Costs is not covered by this Sub-Clause or other Clauses of these Conditions, the Contract Price stated in the Contract Agreement shall be deemed to have included amounts to cover the contingency of other rises and falls in costs.

就本款或本条件其他条文未完全抵补费用任何涨落来说，应认为合同协议书载明的合同价已经列入了足以支付费用中其他涨落的不可预见费。

【译注：to the extent that在这里是限制"合同价列入不可预见费"的程度。"本款或本条件其他条文未完全抵补费用任何涨落"是"合同价列入不可预见费"的原因和范围。

cover的意思与用法，请见第4.11款译注。文献[6]将the Accepted

Contract Amount shall be deemed to have included amounts to cover the contingency of other rises and falls in costs译成了"中标合同金额应被视为已包括其他成本涨落的应急费用。"未将amounts to cover翻译出来。have included（已包括）的宾语是amounts，不是contingency of other rises and falls in costs。】

14.1 The Contract Price 合同价

(b) the Contractor shall pay all taxes, duties and fees required to be paid by the Contractor under the Contract, and the Contract Price shall not be adjusted for any of these costs, except as stated in Sub-Clause 13.6 [Adjustments for Changes in Laws]; and

（b）承包人应支付本合同要求由其支付的所有税金、关税和杂费，并且除了第13.6款[因法律改变的调整]说明的情况以外，合同价不应为这些费用（指所有税金、关税和杂费）任何一种而调整；

【译注：文献[8]错将all taxes, duties and fees笼统译成"各项税费"。

文献[6]、[7]和[8]都错将the Contract Price shall not be adjusted for any of these costs译成了"合同价不应因这些费用进行调整"，会使人以为"这些费用（指所有税金、关税和杂费）"会成为调整合同价的原因。一般来说，所有税金、关税和杂费都是政府或其他公共机构一次索取的，缴纳之后不会再变，因而不会成为调整的原因。】

14.2 Advance Payment 预付款

The Contractor shall ensure that the Advance Payment Guarantee is valid and enforceable until the advance payment has been repaid, but its amount may be progressively reduced by the amount repaid by the Contractor.

承包人应确保该预付款归还保证书在还清预付款之前一直有效

且可执行,但数额可逐步减去承包人归还的数额。

【译注:文献[8]将may be progressively reduced by the amount repaid by the Contractor译成了"可根据承包商付还的金额逐渐减少。"by the amount repaid by the Contractor不是"根据承包商付还的金额"的意思,而是动词reduce要求的,表示减少的量。】

(i) the estimated contract value of the Works executed, and the Contractor's Documents produced, up to the end of the period of payment (including Variations but excluding items described in sub-paragraphs (ii) to (x) below);

(ⅰ)该支付期末之前本工程已实施部分和已做完承包人文件的估算合同价值(包括各项变更,但不包括下文(ii)至(v)项所列之项);

【译注:produce v.t. to cause to have existence or to happen:(做成,做完)[12]这里用的是produce,而不是submit。或许,有些承包人文件不易提交;或许,由于承包人文件是在远处做成,在申请期中款时还来不及交到雇主手里。只要做成了,就可申请付款。文献[8]将produce译成了"提出",值得商榷。】

14.3 Application for Interim Payment 申请期中支付

(vi) any other additions and/or deductions which have become due under the Contract or otherwise, including those under Sub-Clause 3.5 [Agreement or Determination];

(ⅵ)根据本合同或其他理由已到期的任何其他添加和/或扣除金额,包括按第3.5款[商定或决定]商定或决定者;

【译注1:2017年第2版将1999年第1版跟在any other additions or deductions which may have become due under the Contract or otherwise,后面的including those under Clause 20[Claims, Disputes and Arbitration]; and (f) the deduction of amounts included in previous Statements.改成了including those under Sub-Clause 3.5 [Agreement or Determination]。

译注2：文献[8]将any other additions or deductions which may have become due under the Contract or otherwise, including those under Clause 20[*Claims, Disputes and Arbitration*]译成了"根据合同或包括第20条[索赔、争端和仲裁]等其他规定，应付的任何其他增加或减少额；"

FIDIC合同条件参照了适合于普通法的ICE合同条件，所以，承包人不但可以根据合同，也可以根据普通法向雇主要求补偿。而普通法很大一部分是惯例，掌握在法官手里，并无成文的"规定"。这句话中的otherwise的意思就是承包人可根据合同之外的普通法其他部分向雇主提出补偿要求。所以，译文可改成"根据本合同或其他理由，包括第20条[索赔、争端和仲裁]，可能到期应付的任何其他增加或减少额"。】

14.5 Plant and Materials intended for the Works 拟用于工程的装备和材料

(iii) submitted a statement of the Cost of acquiring and shipping or delivering (as the case may be) the Plant and Materials to the Site, supported by satisfactory evidence; and either:

（iii）提交了获取和装运装备和材料或将其送达现场（视情况而定）的费用报表，并附有令人满意的证据；或者

【译注：文献[6]将acquiring译成"购买"，其实，装备和材料也可以赊购、租用，不一定购买。】

15.1 Notice to Correct 纠正通知

If the Contractor fails to carry out any obligation under the Contract the Employer may, by giving a Notice to the Contractor, require the Contractor to make good the failure and to remedy it within a specified time ("Notice to Correct" in these Conditions).

承包人若未履行本合同中任何义务，雇主可发通知给承包人，要求在某具体说明的时间内，赔偿并纠正因此而带来的后果（本条件称"纠正通知"）。

【译注：文献[8]将其译成了"如果承包商未能根据合同履行任何义务，雇主可通知承包商，要求其在规定的合理时间内，纠正并补救上述未履约。"

这里的问题是，如果因承包人未履行合同中任何义务，造成后果之后，承包商应当和能够纠正并补救的是什么？在多数情况下应当是后果，而不是"未履约"本身。make good和remedy 的意思分别是"补偿，赔偿，偿付；实现（意图、诺言）"和"补救；纠正，改善；修补，修缮"。所以，可将"纠正并补救上述未履约"改成"赔偿并纠正因此而带来的后果。"】

15.3 Valuation after Termination for Contractor's Default 因承包人不作为终止后估价

The Employer may withhold payment to the Contractor of the amounts agreed or determined under Sub-Clause 15.3 [Valuation after Termination for Contractor's Default] until all the costs, losses and damages (if any) described in the following provisions of this Sub-Clause have been established.

雇主可到查明本款下文所言全部可能有的费用、损失和赔偿金时，再支付按第15.3款[因承包人不作为终止后估价]商定或决定的数额。

【译注：文献[8]将1999年第1版中的until…have been established 译成了"在确定……前"。将establish译成"确定"不恰当，一是因为容易和从determine和define翻译过来的"确定"混淆；二是establish无"确定"的意思。Establish *v.t.* to cause to be accepted or recognized（证实、查明）[15] 】

15.5 Termination for Employer's Convenience 为雇主方便终止

The Employer shall be entitled to terminate the Contract at any time for the Employer's convenience, by giving a Notice of such termination to the Contractor (which Notice shall state that it is given under this Sub-Clause 15.5).

雇主应有权在任何时候发终止通知给承包人，为雇主的方便而终止本合同（通知应指出是按本15.5款发出的）。

【译注：The Employer shall be entitled to terminate the Contract at any time for the Employer's convenience 在1999年第1版中是 The Employer shall be entitled to terminate the Contract, at any time for the Employer's convenience，文献[8]将其译成了"雇主应有权在对他方便的任何时候，……终止合同"。其实，在at any time前面的逗号是多余的，2017年第2版将其删掉。这样一来，for the Employer's convenience就不是any time的后置定语，而是terminate的状语。】

17.3 Intellectual and Industrial Property Rights 知识与行业产权

【译注：industry有工业、行业、产业的意思，工业是现代国民经济各种行业中的一种。若不问具体背景，就译成"工业"，不恰当。Property原本既有财产，又有财产权、所有权的意思。本条件英文编者可能因担心一些人将其理解局限于"财产"，才添加了Rights一词。】

Whenever a Party receives a claim but fails to give a Notice to the other Party of the claim within 28 days of receiving it, the first Party shall be deemed to have waived any right to indemnity under this Sub-Clause.

当一方收到某索要，但未在此后28天内将其通知另一方，就应认为该方已放弃按本款得到保障的任何权利。

【译注：文献[8]将1999年第1版中的Whenever a Party does not give notice to the other Party of any claim within 28 days of receiving the claim. 译成了"当一方未能在收到任何索赔28天内，向另一方发出关于索赔的通知时"，易使人以为，一方向另一方索赔。】

If a Party is entitled to be indemnified under this Sub-Clause, the indemnifying Party may (at the indemnifying Party's cost) assume overall responsibility for negotiating the settlement of the claim, and/or any litigation or arbitration which may arise from it. The other Party shall, at the request and cost of the indemnifying Party, assist in contesting the claim. This other Party (and the Contractor's Personnel or the Employer's Personnel, as the case may be) shall not make any admission which might be prejudicial to the indemnifying Party, unless the indemnifying Party failed to promptly assume overall responsibility for the conduct of any negotiations, litigation or arbitration after being requested to do so by the other Party.

一方若按本款有权得到保障，则保障方可（由其承担费用）负全责通过谈判解决该项索要，和/或可能因此而引起的任何诉讼或仲裁。在保障方提出请求并承担费用时，另一方应在该索要中协助争辩。除非保障方在另一方提出请求后未立即承担进行任何谈判、诉讼或仲裁的全责，这另一方（以及承包人人员或雇主人员，视情况而定）不应做出任何可能损害保障方的退让。

【译注：文献[8]将This other party (and its Personnel) shall not make any admission which might be prejudicial to the indemnifying Party,译成了："此另一方（及其人员）不应做出可能损害保障方的任何承认，"

admission有"承认（事实、错误）"的意思，若按此义将admission译成"承认"，声称"公正"的FIDIC会鼓励合同双方不承认事实吗？显然不会。实际上admission还有"允许他人进入"的意思,另外,再设想一下保障方在与索要者谈判或争辩中的情景,无非就是双方进攻、退守的较量。contest就是攻或以攻为守,而admission就是退或让。所以,可以将admission译成"退让"、"给对方留空子"或"让对方有机可乘"。】

19.1 General Requirements 一般要求

Without limiting either Party's obligations or responsibilities under the Contract, the Contractor shall effect and maintain all insurances for which the Contractor is responsible with insurers and in terms, both of which shall be subject to consent by the Employer. These terms shall be consistent with terms (if any) agreed by both Parties before the date that both Parties signed the Contract Agreement.

在不限制任何一方本合同义务或责任的条件下,承包人应按雇主应允的条款从其应允的保险人处取得由其负责的所有各项保险。这些条款应同签署合同协议书日前双方可能商定的条款一致。

【译注:文献[6]至[8]将These terms shall be consistent with any terms agreed by both Parties before the date …中的terms译成了"条件"。实际上,terms中不但有conditions(条件)还有warranties(保证条文)。所以,译成"条件",就丢失了warranties的含义。请见下文。

terms of a contract 合同条款

All the obligations and rights agreed between the parties and all the terms implied by law. In English contracts the terms may consist of conditions and warranties. However, it should be noted that the phrase 'terms and conditions' is commonly used in contracts. This phrase is used in order to convey that there are some terms in the contract (conditions) which are considered to be more important than others. 合同双

方同意的全部义务和权利，以及法律隐含的所有条款。在英国的合同中，这些条款可以由基本条款和保证条款组成。然而，应该注意的是，合同中常用"条款和条件"一语。使用该用语的目的是要说明合同中有些条款（基本条款）要比其他条款更重要。】

If the Contractor fails to effect and keep in force any of the insurances required under Sub-Clause 19.2 [Insurances to be provided by the Contractor] then, and in any such case, the Employer may effect and keep in force such insurances and pay any premium as may be necessary and recover the same from the Contractor from time to time by deducting the amount(s) so paid from any moneys due to the Contractor or otherwise recover the same as a debt from the Contractor. The provisions of Clause 20 [Employer's and Contractor's Claims] shall not apply to this Sub-Clause.

承包人若未投保第19.2款[承包人投保]要求的任何事项并使之持续有效，那么，雇主可在任何此种情况下，为上述各险投保并使之持续有效，并缴纳任何可能必要的保险费，然后，可从任何应付承包人的款项中分次扣除如上支付的各数额，随时将其从承包人处收回，亦可另外当作一债款从承包人处讨回。第20条[雇主与承包人索要]条文不适用于本款。

【译注：insurance n.原意是(Undertaking, by a company, society, or the State to provide)safeguard against loss, provision against sickness, death, etc., in return for regular payments.

Insure v.t. make a contract that promises to pay, secures payment of, a sum of money in case of a accident, damage, loss, injury, death etc.

清代魏源，在《海国图志》中将insurance（insure）译成"担保"。"(英吉利国）广推贸易之法，有火轮船，航河驶海，不待风水，又造辘路用火车，往来一时可行百有八十里，虞船货之存失不定，则又约人担保之，设使其船平安抵岸，每银百两给保价三四元,即如担保一船二万银,则预出银八百元,船不幸沉沦,则保人给偿船主银二万两。"

日本福泽喻吉译为"保险",传到我国,沿用至今。显然,魏源的"担保"比"保险"更明白地反映了insurance(insure)的原意。"保险"中的"保",有保护、保卫;保持;保证、承诺做到之意。"险",有危险、险要去处之意。按汉语习惯,"保险"就是"保护危险"、"保卫危险"、"保持危险"、"保证危险"、"承诺一定有危险",等的意思,同insurance(insure)的原意岂止相差十万八千里,简直是背道而驰。

insurance(insure)的原意,从投保人来说,实际是"避险"、"防险"、"御险";从保险人来说,是许诺若有不测,兑现补偿投保人损失、损坏或伤害的诺言。

既然如此,按理,"保险"应弃而不用。无奈,它已经深深进入国人思维之中,难以动摇。另外,"担保"已经用于security和collateral的汉译之中,不易再改。既然国人已经习惯,也大致明白了"保险"的真正含义,维持现状也行。不过,当insurance(insure)与其他英文词语一起使用时,若译成汉语,就要考虑"保险"这个词的弊病。有时,需要将其分成两个字,单独使用。】

19.2 Insurance to be provided by the Contractor 承包人投保

The Contractor shall provide the following insurances:
承包人应为如下各项投保:

【译注:provide在这里是"设立",不是"提供"之意。《现代汉语词典》:提供:供给(意见、资料、物资、条件等);《新华字典》:提供:供给可参考或利用的(意见、资料、物资、条件、情况等)。上述释义暗含"供给"的是确定,而非或然之物。但是,保险给受益人(除了与雇主联名投保,受益人就是承包人自己)带来的却是或然之物:只有遇到不测时才能得到补偿。另一方面,既然是"供给",就有供给者与接受者。若将provide the following insurances译成"提供如下各项保险",那么,谁是"各项保险"的接受者呢,接受者拿到"各项保险"后如何"参考或利用"呢?

provide insurances也可译成"投保"。承包人投保,就是为了自己可能遇到的危险设防,预设防备之策,不是为了提供他人。

还有,将provide the following insurances译成"提供如下各项保险"与第19.1款矛盾:the Contractor shall effect and maintain all insurances for which the Contractor is responsible with insurers and in terms, both of which shall be subject to consent by the Employer.(承包人应按雇主应允的条款从其应允的保险人处取得由其负责的所有各项保险。)"承包人从保险人处取得的保险""提供"给谁呢?承包人为何要将自己可能受益的东西"提供"给别人呢?】

19.2.1 The Works 本工程

(a) the Works and Contractor's Documents, together with Materials and Plant for incorporation in the Works, for their full replacement value.

The insurance cover shall extend to include loss and damage of any part of the Works as a consequence of failure of elements defectively designed or constructed with defective material or workmanship; and

(a)本工程与承包人文件,连同用在本工程之中的材料和装备的完全重置费用。保险额还应足以抵补本工程任何部分因设计或施工使用了有缺陷的材料或工艺而使设计或施工不当的组件失效造成的损失与损坏;

【译注:insurance cover可译成"保险额"。extend v.i. 是在某基数之外"扩"、"展"、"增"、"添",等之意。因而,insurance cover shall extend to include译成"投保额还应足以……"。】

The insurance cover shall cover the Employer and the

Contractor against all loss or damage from whatever cause arising until the issue of the Taking-Over Certificate for the Works. Thereafter, the insurance shall continue until the date of the issue of the Performance Certificate in respect of any incomplete work for loss or damage arising from any cause occurring before the date of the issue of the Taking-Over Certificate for the Works, and for any loss or damage occasioned by the Contractor in the course of any operation carried out by the Contractor for the purpose of complying with the Contractor's obligations under Clause 11 [Defects after Taking Over] and Clause 12 [Tests after Completion].

保险额应足以使雇主与承包人防备颁发本工程接收证书之前因各种原因造成的所有损失或损害。此后，任何未完工作在颁发本工程接收证书前因任何原因造成的损失或损害，以及承包人在为遵守第11条[接收后缺陷]和第12条[完工后检验]中的义务而进行的任何作业过程中引起的任何损失或损坏的保险应继续到颁发完工证书之日。

【译注：shall cover里的cover是动词，意思是"（钱）足以……"或"弥补"。】

21.4.1 Reference of a Dispute to the DAAB 将争议提交避免/裁决争议小组

The arbitrator(s) shall have full power to open up, review and revise any certificate, determination (other than a final and binding determination), instruction, opinion or valuation of the Employer and/or of the Employer's Representative, and any decision of the DAAB (other than a final and binding decision) relevant to the Dispute.

仲裁人应有充分权利拆阅、审核和修改雇主和/或雇主代表就该争议发出的任何证书、决定（最后并有约束力者除外）、指示、意见或估价，以及DAAB对此争议的任何裁定（最后并有约束力者除外）。

【译注：文献[4]至[8]都将open up译成"公开"。open up是"启封、拆阅（可能已封存之物）"的意思。仲裁人没有必要"公开"，对谁"公开"呢？】

If the Employer anticipates that a Subcontractor is to be instructed under Sub-Clause 13.3 [Variation Procedure] but is not to be a nominated Subcontractor this Sub-Clause should be amended, describing the particular circumstances.

如果雇主预计要按第13.4款[变更程序]指示某分包人，但不是指示为指定分包人，本款应当修改，说明这些具体情况。

【译注：文献[6]将红皮书1999年第1版中的If the Employer anticipates that a Subcontractor is to be instructed under Sub-Clause 13 but is not to be a nominated Subcontractor译成了"如果雇主预期分包商按照第13条接受指示，但不算指定的分包商"。承包人的分包人有两种，由雇主或雇主代表指定的和承包人自己选定的（domestic sub-contractors）。当雇主指示承包人变更的时候，承包人可能不愿意，或者无力进行。雇主就可能指示某家承包人选定的分包人实施该变更。】

（四）Advisory Notes to Users of FIDIC Contracts Where the Project Uses Building Information Modelling Systems 对项目使用建筑信息模型的FIDIC合同用户的若干建议说明

BIM is not a set of contract conditions; it is a mechanism to provide an environment where all parties have access to information relevant to their role in the design and construction of a project. Wherever possible, a combined (sometimes called federated or collaborative) model is developed for all parties to share, even if, as is often the case, various designers have used different computer aided design programs to develop their respective designs. Drawings and specifications are held in a common database accessible to everyone who can be used for clash detection, coordination of designs, communication of changes, and construction sequencing.

BIM不是一套合同条件，而是一种机制，构筑所有参与者都可访问同其在项目设计与施工中的角色有关的信息的环境。各种不同设计人即使经常用不同的计算机辅助设计程序逐步完善各自的设计，凡有可能之处，也会生成一种所有参与者交流共用的组合模型（有时称为联盟或合作模型）。图样与设计说明书保存在一个共用数据库之中，供每个人访问，每个人都可充当检测冲突、协调设计、通知修改，以及安排施工顺序之用。

【译注：文献[16]将it is a mechanism to provide an environment where all parties have access to information relevant to their role in the design and construction of a project.译成了"而是提供了一个平台，通过这个平台所有项目相关方可接触到与其设计与施工任务相关的信息。"

provide *v.t.* to afford or yield（出产；产生）[12]】

Coordination of goals and effort is essential and is generally achieved by a BIM Protocol and a BIM Execution Plan, both key documents to access and understand work in this environment. A designer needs to understand and work to the Levels of Design (or Detail) (LOD) that will be spelled out in these documents to ensure that there is sufficient detail at each level to allow all designs to progress efficiently and avoid unnecessary changes.

目标与努力的统一至关重要，一般是利用BIM协议和BIM实施计划实现的，这两个都是要访问和理解在该环境中工作的重要文件。设计者需要理解这两个文件中力图表达的设计（或大样）的深度（LOD）（要求）并努力达到，确保每一深度都足够详细，使所有的设计都进展省时、节费，避免无必要的变更。

【译注：文献[16]将A designer needs to understand and work to the Levels of Design (or Detail) (LOD) that will be spelled out in these documents to ensure that there is sufficient detail at each level译成了"设计人员需要理解并遵从从BIM协议和BIM执行计划中说明的多设计（或细节）层次（LOD），确保每个层次都包含足够的细节"。

设计的不同阶段，对设计文件深度的要求不同。很多设计人员常常难以理解，更难做到恰到好处。对设计文件的深度提出清楚明白的书面要求，亦非易事。所以，用了spell out来表示BIM协议和BIM执行计划的这一难言之苦。Spell *v.t.* make out(words, writing) laboriously, slowly（（通常与out连用）费力而缓慢地读懂）。Levels of Design (or Detail) (LOD)是"设计（或大样）的深度（LOD）"的意思。】

This involves not only the drawing elements but also the embedded data. Experience shows that this is a significant effort, so responsibilities for completing this task should be clear and appropriate allowances provided.

这不仅仅涉及图样的各个组成部分，也涉及写入图样中的数据。经验表明，这是一种非同寻常的努力，因而，完成这一任务的责任应当清楚明白，并应当留有适当的余地。

【译注：文献[16]将这一句译成了"包括图纸和嵌入的数据，经验表明，这项工作很重要，因此完成此项工作的责任应清晰并且应提供适当的酬劳。"

Allowance *n. sum of money, amount of sth., allowed to sb.*（津贴；特别经费；所允许给予之量）[10]。津贴，固然也是一种酬劳，但不是酬劳主要部分。另外，在FIDIC这个文件里，用remuneration表示给予DAAB成员、调解员、投标人的酬劳（Employers should also consider remunerating tenderers if, in order to provide a responsive Tender, they have to undertake studies or carry out preliminary design work.（雇主还应考虑在投标人为了编制出符合招标要求的投标文件而不得不承担研究或进行初步设计工作时，给予报酬的问题。）所以，文献[16]将allowances译成"酬劳"，很值得再想一想。其实，这里的allowances是*amount of sth.*的意思。上面这句话里虽然说so responsibilities for completing this task should be clear（完成这一任务的责任应当清楚明白），但是，这一任务的责任不容易弄清楚明白，所以，就退一步，and appropriate allowances provided（应当留有适当的余地），appropriate allowances provided是appropriate allowances should be provided的简化，其中provide不是"提供"的意思，是"设置、设立"的意思。】

Legal counsel should review the contract to ensure that it does not create an unintended joint venture which may be a risk in some jurisdictions.

法律顾问应当审核合同，确保在某些法律管辖区不使其成为会是风险，但并非本意的共同冒险行动。

【译注：文献[16]将joint venture译成"联营体"。joint venture在这里是共同冒险的意思，不是"联营体"。Venture *n.* undertaking in which there is risk.（冒险；冒险事业）[10]】

（五）NOTES ON THE PREPARATION OF SPECIAL PROVISIONS 编写特殊条文应注意之处

Sub-Clause 5.2 Contractor's Documents 承包人文件

- making necessary connections to the Plant.
- 接通与该装备连接的必要管线。

【译注：文献[4]将这句话译成了"制作生产设备必要的连结。"在这里动词make没有实际含义，不是"制作"的意思。】

Sub-Clause 14.1 The Contract Price 合同价

Normally, an EPC/turnkey contract is based on a lump sum price, with little or no measurement.

一般情况下，EPC/交钥匙合同是按总价支付的，很少或根本不计量工程量。

【译注：建筑、土木、机电，以及其他工程的施工合同、设计施工合同，甚至某些EPC合同，在签约前、后，在设计、招标、施工等阶段，都要计量工程数量。英文的"计量"有三个词：measurement、remeasurement和admeasurement。这三个词的区别，可见如下一段话：This paragraph states that the rules and provisions used in the pre-contract exercise of measuring the work also apply to the post-contract task of measurement. The correct term for this task is re-measurement where the work is physically measured on site or admeasurement where the actual quantities are calculated from records. 本段说明，签订合同前计量各个分项的规则和规定也适合于签订合同之后计量各个分项。签订合同后计量各个分项这一任务正确名称，当在现场实际计量时叫作"实量"（re-measurement），而当根据记录计算时叫作"推量"（admeasurement）。[14] 不少人不太明白re-measurement的真实含义，按字面译成了"重新计量"。这样翻译，人们就会问，"原来计量"是什么人在何时做的？大概是FIDIC为了少费口舌，便在2017年第2版将1999年第1版的remeasurement改成了measurement。】

Additional Sub-Clauses may be required to cover any exceptions to the provisions set out in Sub-Clause 14.1, and any other matters relating to payment.

可能需要添加若干款，用来处理第14.1款已列出条文的任何例外，以及与支付有关的任何其他事项。

【译注：文献[4]将1999年第1版的Additional Sub-Clauses may be required to cover any exceptions to the options set out in Sub-Clause 14.1, and any other matters relating to payment. 译成了"为了包括第14.1款中所列可选内容以外的任何其他内容，以及有关付款的任何其他事项。可能需要一些附加条款。"

这样的译文使人觉得"一些附加条款"的目的就是"包括""第

14.1款中……其他内容,以及……其他事项。"其实,此处Cover不是"包括"的意思,而是"处理"。Cover v.t&i. deal with(处理)[12]。】

(六)Annex A EXAMPLE FORM OF PARENT COMPANY GUARANTEE 母公司保证书格式样本

In consideration of you, the Employer, awarding the Contract to the Contractor, we (name of parent company) irrevocably and unconditionally guarantee to you, as a primary obligation, the due performance of all the Contractor's obligations and liabilities under the Contract, including the Contractor's compliance with all its terms and conditions according to their true intent and meaning.

为报答贵方,雇主,将本合同授予承包人,我方(母公司名称)向贵方无条件、绝不改悔地保证,我方基本义务就是恰当地履行本合同承包人所有义务与责负,包括承包人按所有条款与条件的本意与含义遵守之,

【译注:文献[6]至[8]都将In consideration of 译成了"考虑到"。诚然,字典上in consideration of有"由于,考虑到"之条目[14],然而也有"作为对……的酬报"之解释。但不能译成"考虑到",原因可见下文合同协议书的译注。】

(七)CONTRACT AGREEMENT 合同协议书

Whereas the Employer desires that the Works known as _____ [name and number of the Contract] should be executed by the Contractor, and has accepted a Tender by the Contractor for the execution and completion of these Works and the remedying of any defects therein, for the lump sum Contract Price of:

鉴于雇主欲由承包人实施名为_____的本工程,并已接受了承包人为该工程的实施、完成以及修补其任何缺陷提交的投标文件,总额合同价为:

3. In consideration of the payments to be made by the Employer to the Contractor as hereinafter mentioned, the Contractor hereby covenants with the Employer to design, execute and complete the Works and remedy any defects therein, in conformity with the provisions of the Contract.

3. 承包人特此雇主立约按本合同条文设计、实施和完成本工程，并修补其中任何缺陷，报答由雇主付给承包人下文提及的各款项。

4. The Employer hereby covenants to pay the Contractor, in consideration of the design, execution and completion of the Works and the remedying of defects therein, the final Contract Price at the times and in the manner prescribed by the Contract.

4. 雇主特此立约，保证按本合同规定的时间和方式向承包人支付最终合同价，报答承包人设计、实施、完成本工程并修补其中任何缺陷。

【译注：文献[6]至[8]均将In consideration of译成了"鉴于"。诚然，字典上in consideration of有"由于，考虑到"之条目，然而也有"作为对……的酬报"之解释。那么，为什么应译成"作为对……的酬报"，而不应译为"鉴于"呢？

第一，consideration有"报酬，约因，对价"之含义；

第二，请注意，in consideration of现在是用于合同协议书，而不是别处。众所周知，要约、承诺和报酬是有效合同的三大要素，合同中不能不载明报酬。

第三，表示"鉴于"的意思，原文在这同一文件中用的是"Whereas"。】

In Witness whereof the parties hereto have caused this Agreement to be executed the day and year first before written in accordance with their respective laws.

为见证上述事项，合同双方于上面日期，根据各自的法律签署订立本协议书。

【译注：文献[6]至[8]均将the parties hereto have caused this Agreement to be executed the day中的execute译成了"实施"。】

三　结束语

FIDIC合同格式以往汉译本中问题，同其他英文作品的汉译本一样，原因主要有三：

一、未理解英文原著；

二、汉语表达能力不够；

三、不熟悉设计、施工、制造、贸易、国际工程采购通行做法和合同文件。

此外，现有的英汉词典有缺陷和不足，例如释义不全、用法例句不多、陈旧、编撰者知识面有限，等等。我们在翻译时，首先应从头至尾，通读词典中的所有释义。若无恰当适用者，应查阅英美各国原版词典，当然，也应当充分利用搜索引擎在各个网站上搜索。

对于时人，往往是汉语绊住了译者的手脚，因此，扩充自己的汉语知识，提高使用能力必不可少。

当然，译者疏忽、不经心，也是重要原因。因此，加强社会责任心更是译者的首要任务。

参考文献

[1] 卢谦,译. 土木工程施工国际通用合同条件(附投标书及协议格式)汉英对照[M]. 北京:中国建筑工业出版社,1986.

[2] 卢有杰. FIDIC合同条件理解与译法商榷[J]. 建筑经济,1996(1).

[3] 卢有杰. 关于1999年版FIDIC合同条件的理解与翻译[J]. 工程经济,2011(8);2011(9).

[4] 国际咨询工程师联合会,编. 土木工程施工合同条件应用指南[M]. (1987年第4版、1988年订正版). 北京:航空工业出版社,1991年4月第1版.

[5] 国际咨询工程师联合会,编. 设计-建造与交钥匙工程合同条件(1995年第1版)[M]. 北京:中国建筑工业出版社,1996年11月第1版.

[6] 国际咨询工程师联合会,中国工程咨询协会,编译. 施工合同条件1999年第1版[M]. 北京:机械工业出版社,2002年5月第1版,2010年9月第1版.

[7] 国际咨询工程师联合会,中国工程咨询协会,编译. 生产设备和设计-施工合同条件1999年第1版[M]. 北京:机械工业出版社,2002年5月第1版,2010年9月第1版.

[8] 国际咨询工程师联合会,中国工程咨询协会,编译. 设计采购施工(EPC)/交钥匙工程合同条件1999年第1版[M]. 北京:机械工业出版社,2002年5月第1版,2010年9月第1版.

[9] 国际咨询工程师联合会,中国工程咨询协会,编译. 菲迪克(FIDIC)合同指南[M]. 北京:机械工业出版社,2003年6月第1版.

[10] World Bank. *Standard Bidding Documents-Procurement for Works*. May 2000.

[11] World Bank. *Standard Bidding Documents-Procurement of Works and User's Guide*. May 2005.

[12] Merriam-Webster Inc. *Merriam-Webster's Collegiate Dictionary*, Tenth Edition. Merriam-Webster Inc. (1997).

[13] Oxford University Press. The Advanced Learner's Dictionary of Current English with Chinese Translation(现代高级英汉双解辞典),1982年第14版.

[14] 上海译文出版社,新英汉词典(增补本),上海:1985年7月,新2版.

[15] Random House Inc., The Random House Dictionary, New York, 1980.

[16] 陈勇强、吕文学、张水波,著. FIDIC 2017版系列合同条件解析[M]. 北京:中国建筑工业出版社,2019年4月第1版.